The Bose-Einstein condensation of electrons: a long overdue discovery of how superconductors really work

András Kovács, Giorgio Vassallo, Paul O'Hara, and Chris Scott

PUBLISHED BY THE ZITTER INSTITUTE (ZITTER-INSTITUTE.ORG). THE BOOK CONTENT DOES NOT NECESSARILY REPRESENT A CONSENSUS OF ALL ZITTER INSTITUTE MEMBERS.

ISBN 978-952-65314-4-1 (SOFTCOVER)

ISBN 978-952-65314-5-8 (PDF)

Contents

Foreword

The discovery of superconductivity is over 100 years old. While superconducting materials have been studied in much detail over the past 100 years, it remains a grand intellectual challenge to understand how metals completely loose electric resistivity at low temperatures, and it remains an open question whether room temperature superconductivity can be achieved. Superconductivity based technologies are gradually being commercialized, in powerful magnetic field generators, SQUID magnetometers, Josephson junction amplifiers, microwave filters, and superconducting qubits. The development of these practical technologies would also benefit from understanding the **real** physics of superconductivity. Inspired by these challenges, we believe it is time to properly understand how superconductivity really works.

The vast collection of superconductivity related measurement data, accumulated over the past 100 years, reveals quantitative formulas that characterize superconductivity. The most important such formulas are the London equation, the London moment formula, the Josephson frequency formula, the Uemura scaling, and the Roeser–Huber formula. A correct theory of superconductivity must rigorously derive all of these formulas - without involving ad-hoc assumptions or fitting parameters. While the currently favored theory of superconductivity assumes the existence of free-flowing electron pairs whose kinetic energy is near the Fermi energy level, the essential superconductivity phenomena are in fact incompatible with such free-flowing electron pairs. If the Meissner effect was caused by electron pairs freely circulating around the perimeter, these radially accelerating electrons would loose energy by emitting radiation; in contrast, the London equation formula is static. If the magnetic field of rotating superconductors was induced by freely circulating electron pairs, the London moment formula would depend on number of involved electron pairs; in contrast, the London moment formula contains neither the number or density of superconducting electrons. If the Josephson radiation was caused by such freely oscillating electron pairs, the derivation of Josephson frequency formula would not be based on electrons having close to zero kinetic energy. If such free-flowing electron pairs were involved in high-temperature superconductivity, the currently favored models would have predicted the experimentally observed Uemura scaling of superconducting temperature. These paradoxes demonstrate that the currently favored BCS theory of superconductivity is fundamentally wrong, and thus shall never be able to predict higher temperature superconductors. Not surprisingly, the BCS theory failed to predict any improved superconductor for 60 years already; cuprates, MgB_2, or iron-based superconductors were all discovered by trial and error.

In this book, we develop the theory of electrons' Bose-Einstein condensation. The key is to correctly calculate both the microscopic interaction and the macroscopic thermodynamics of Bose-Einstein condensed electrons. Recognizing that only a fraction of electrons condenses at the transition point, a realistic calculation of Bose-Einstein condensation temperature requires evaluating the thermodynamic balance between coherent and incoherent electron populations. All results are derived through mathematically rigorous calculations that are explained step by step.

After clarifying the dynamics of Bose-Einstein condensed electrons from first principles, the essential formulas of superconducting materials emerge naturally: this book contains rigorous derivations of the London equation, the London moment formula, the Josephson frequency formula, the Uemura scaling, and the Roeser–Huber formula. For most of these formulas, our book contains their first mathematically and physically correct derivation. We also review a large number of experiments that show direct signatures of Bose-Einstein condensation, including the well-known coherence of superconducting electrons.

Regarding the mystery of superconductors' perfect conductivity: superconducting electrons must have the ability to pass through the lattice without any scattering. This is naturally achieved by the main feature of Bose-Einstein condensed electrons: adding and removing the lowest-energy conduction band electrons whose wavefunction has macroscopically large wavelength - such a large wavelength no longer scatters on defect sites or lattice distortions because of the many orders of magnitude mismatch with respect to the inter-nuclear distance. Regarding the mystery of superconductors' perfect diamagnetism: we show that Meissner flows are generated by coherent electron oscillations that become energetically favored over the non-oscillating state - similarly to the operating principle of free electron lasers.

While Bose-Einstein condensed electron states generally arise in the conduction band, nothing in the theory is specific to just that electron population. This leads to the question: can ordinary electron orbitals host Bose-Einstein condensed electrons? The last chapter presents experimental investigations of this question, with surprising answers.

At the end of this short journey, the reader is rewarded by understanding how superconductivity really works, and learns about unexpected new materials comprising Bose-Einstein condensed electron orbitals.

DEFINITION. A **metal** is defined as a material comprising "conduction band" electrons that are not bound to any specific nucleus. A conduction band comprises anti-bonding orbitals, and the relative energy levels of such orbitals determine whether any electrons occupy them. Since conduction band electrons are not bound to any individual nucleus, they are bound by the positive charge arising on the surface of metallic materials. Therefore the energy eigenstates of conduction band electrons are the standing wave solutions to the square potential-well problem. Metals conduct electricity through their ability to add and remove conduction band electrons at the lowest unoccupied wavelength of such standing wave solutions. These lowest energy unoccupied standing waves have comparable wavelength to the inter-nuclear distance. Electric resistivity arises at the crystal defect sites, as a consequence of a scattering process: the wavefunction of a current-carrying electron scatters on a defect site. The presence of crystal defects is unavoidable in any macroscopic material.

A **superconductor** is a sub-class of metals, defined by its ability to conduct electric current without any measurable resistance. Electric resistivity is caused by scattering on crystal defect sites, lattice vibrations, electron-electron scattering, etc. Superconducting electrons must have the ability to pass through the superconducting material without any such microscopic-scale impact. The simplest way to achieve this is to add and remove those conduction band electrons whose wavefunction has macroscopically large wavelength. Such a large wavelength no longer scatters on defect sites or lattice distortions because of the many orders of magnitude mismatch with respect to the inter-nuclear distance. The principle of Occam's razor dictates that superconductivity theory should focus on models that imply variable occupancy level of the lowest wavelength standing waves. We adhere to this principle in our work.

CHAPTER 1

Spin correlations and electron coherence

Giorgio Vassallo[1,2], **Paul O'Hara**[3], and **Andras Kovacs**[4]
[1] International Society for Condensed Matter Nuclear Science (ISCMNS)-UK.
[2] Università degli Studi di Palermo - Engineering Department, Palermo, Italy.
[3] Istituto Universitario Sophia, Figline e Incisa Valdarno (FI), Italy.
[4] ExaFuse
E-mail: giorgio.vassallo@unipa.it

1.1. A brief history of electron statistics rules

Until the dawn of quantum mechanics, there was little overlap between physics and chemistry. This situation changed dramatically with the introduction of quantum mechanics: the accurate calculation of electron binding energies has become quantum mechanics' main success, thus being able to clarify why fluorine takes an electron from lithium, and not vice versa. This capability of binding energy calculations has eventually established quantum mechanics as a practical tool for the predictive modeling of chemical reactions.

However, the calculation of electron binding energies explains only half of chemistry; the remaining challenge 100 years ago was to explain the phenomenological observation that only two electrons can occupy any given atomic or molecular orbital. This electron occupancy limiting rule has become known as the exclusion principle. Those quantum mechanical systems where the exclusion principle applies are said to obey Fermi-Dirac statistics, while quantum mechanical systems without the exclusion principle are said to obey Bose-Einstein statistics.

The initial formulation of electron statistics rules, dating back to the first half of 20th century, postulated that i) individual electrons cannot occupy the same quantum mechanical state, and ii) electron pairs cannot occupy the same quantum mechanical state. The second postulate refers to the observation that bound electron pairs occupy distinct orbitals, i.e. they generally do not collapse down into the lowest-energy K-shell.

This set of electron statistics postulates was reformulated in the mid-20th century by superconductivity researchers, postulating that i) individual electrons cannot occupy the same quantum mechanical state, ii) hypothetical phonon-bound electron pairs can occupy the same quantum mechanical state, and iii) other electron pairs cannot occupy the same quantum mechanical state, thus maintaining the Fermi-sea of conduction band electrons and the distinct inner-shell electron pairs.

After the discovery of high-temperature superconductivity in the late 20th century, where phonons play no role, the set of electron statistics postulates was again reformulated, postulating that i) individual electrons cannot occupy the same quantum mechanical state, ii) hypothetical phonon-bound electron pairs can occupy the same quantum mechanical state, iii) the hypothetical weakly bound electron pairs of high-temperature superconductors can occupy the same quantum mechanical state **[1]**, and iv) other electron pairs cannot occupy the same quantum mechanical state.

The recently discovered "spin-triplet superconductivity" represents an other major extension of electron statistics postulates; it allows two individual electrons to occupy the same quantum mechanical state as well [2]. The study of various conduction band electron topologies also led to an extension of electron statistics postulates for the case of 2-dimensional electron systems; the scientific literature refers to this new postulate as the "anyon" configuration of electrons [3].

The above-outlined historic trend of continuously extending electron statistics postulates demonstrates that this topic has not been well understood in the first place, thus necessitating additional postulates to accommodate each new experimental discovery. This brings into question what really determines whether electron interactions are fermionic or bosonic. In this chapter, we explore a new understanding of electron statistics, and lay the foundations for understanding electrons' Bose-Einstein condensation.

1.2. Coherent versus incoherent electron states

Let us introduce the following classification of electron states:

- N electrons are said to be in **coherent** state if all quantum numbers of each electron are the same: i.e. they are all in the same quantum mechanical state.
- N electrons are said to be in **incoherent** state if each electron is in a different quantum mechanical state.

In a hypothetical coherent state, each involved electron is an indistinguishable part of exactly the same quantum mechanical wave. The N electrons occupying a coherent state comprise a single wavefunction: $\psi = \sqrt{n}e^{i\varphi}\left|s\right\rangle$, where φ is the quantum mechanical phase of the common wavefunction, $n \equiv \frac{N}{V}$ is its electron density, and $\left|s\right\rangle$ is its spin state. In such a coherent state, individual electron properties may be measured only by such methods that resolve much shorter distances than the quantum mechanical wavelength: for example the Compton scattering of radiation with femtometer-scale wavelength still happens from individual electrons. But any lower resolution measurement on coherent electrons is a simultaneous measurement on all involved electrons. I.e. a spin state measurement yields $\left|+\right\rangle$ or $\left|-\right\rangle$ state of the common wavefunction. It is not possible to measure an individual electron's spin state without breaking up the coherent state.

In an incoherent state, each involved electron comprises a different quantum mechanical wave. Principally, it is possible to measure the individual properties of each incoherent electron. I.e. the$\left|+\right\rangle$ or $\left|-\right\rangle$ spin state of individual electrons can be determined by spin state measurements. In the case of interacting electrons, their measured spin values become statistically correlated. Such statistical correlation is referred to as the spin-correlation between incoherent electrons' spin values. The spin-correlation is said to be isotropic is its statistics does not depend on the spatial direction of spin measurement.

In the following, we shall explore the applicable electron statistics for the above-defined incoherent versus coherent state classes.

1.3. Spin correlations between particles occupying different orbitals

Isotropic spin-correlation (ISC) is observed not only for electrons sharing the same orbital, but also for electrons in different orbitals. For example, the ground state oxygen molecule is the so-called "triplet oxygen": it has two electrons occupying two distinct π^* orbitals (anti-bonding orbital), and these two electrons are isotropically spin-correlated into parallel direction.

Isotropic spin-correlation is observed also between the nuclei comprising a molecule. For example, the hydrogen molecule has two nuclear spin isomers: the two protons of "ortho-hydrogen" are isotropically spin-correlated into parallel direction, while the two protons of "para-hydrogen" are isotropically spin-correlated into opposite direction. This system is significant because the two nuclei are well separated in space; the individual spin state of each proton is thus principally measurable.

Isotropic spin-correlation is observed between a bound electron and a nucleus: such interaction generates the hyperfine split of the electron's binding energy. Isotropic spin-correlation is also observed between a delocalized unpaired electron and a nucleus: such an electron-nucleus spin interaction shows up in NMR measurements, contributing to the Knight shift of metals.

There is a simple reason for ISC particle states being ubiquitous: each quantum mechanical state is characterized by a well-defined energy level. This well-defined energy level condition requires isotropic spin-correlation among the particles occupying a given state; otherwise the spin interaction energy would depend on the spin measurement direction, and there would not be a well-defined energy level. The universal presence of ISC electron states thus follows from the well-defined energy condition, and the challenge is to understand what limits the number of ISC electrons occupying a given quantum mechanical state.

In all observations, **ISC always occurs pair-wise**: i.e. the ISC of $N>2$ particles is never observed. Table 1 illustrates this effect for the simplest atoms: an electron is either spin-correlated to an other electron or to a nucleus, but never to both at the same time. Whether we look at particles sharing the same orbital or particles occupying different orbitals, we thus observe exactly the same phenomenology of strictly pair-wise ISC coupling. This suggests the same origin of the ISC coupling limit, and we therefore look for a unifying principle. Taking the example of hydrogen spin isomers, it is obvious that Pauli's microcausality arguments do not apply to well-separated nuclei, and it is also obvious that there would be nothing anti-symmetric about the exchange of two separated nuclei.

TABLE 1. The pattern of spin correlations between electrons and the nucleus

	$p^+ + e^-$ (H)	$p^+ + 2e^-$ (H$^-$)	$^3He^{2+} + e^-$ (He$^+$)	$^3He^{2+} + 2e^-$ (He)
Hyperfine split	yes	no	yes	no
ISC electrons	-	yes	-	yes

We approach the challenge of finding a unifying principle by firstly considering what spin measurement really means. It is well-known by the operators of electron spin resonance (ESR) and nuclear magnetic resonance (NMR) equipment that the particle whose spin they measure is in a Larmor spin-precession. Specifically, the magnetic moment vector generated by the particle's spin is Larmor precessing around the externally applied magnetic field lines. The torque is this Larmor precession is given by:

$$\tau = |g\boldsymbol{\mu}_B \times \boldsymbol{B}| \tag{1.3.1}$$

where B is the applied magnetic field strength, μ_B is the Bohr magneton, and $g \approx 1 + \frac{\alpha}{2\pi}$ is the gyromagnetic ratio of the electron spin.

This Larmor spin-precession's angular frequency calculation can be found in reference **[10]**, and it evaluates to:

$$\omega_L = \frac{g\mu_B}{\hbar}B$$

We note that the Larmor spin-precession's frequency is exactly half of the "electron spin resonance" frequency [**6**]: $\omega_{esr} = 2\omega_L$. The ω_{esr} value is the experimentally measurable angular frequency of resonant flipping between the parallel and anti-parallel spin precession orientations with respect to a sinusoidally varying applied magnetic field. A consequence of Larmor spin-precession phenomenon: the measured $\frac{\hbar}{2}$ spin angular moment, measured via an applied B field, is only that component of the total angular momentum vector which is pointing along the applied B field.

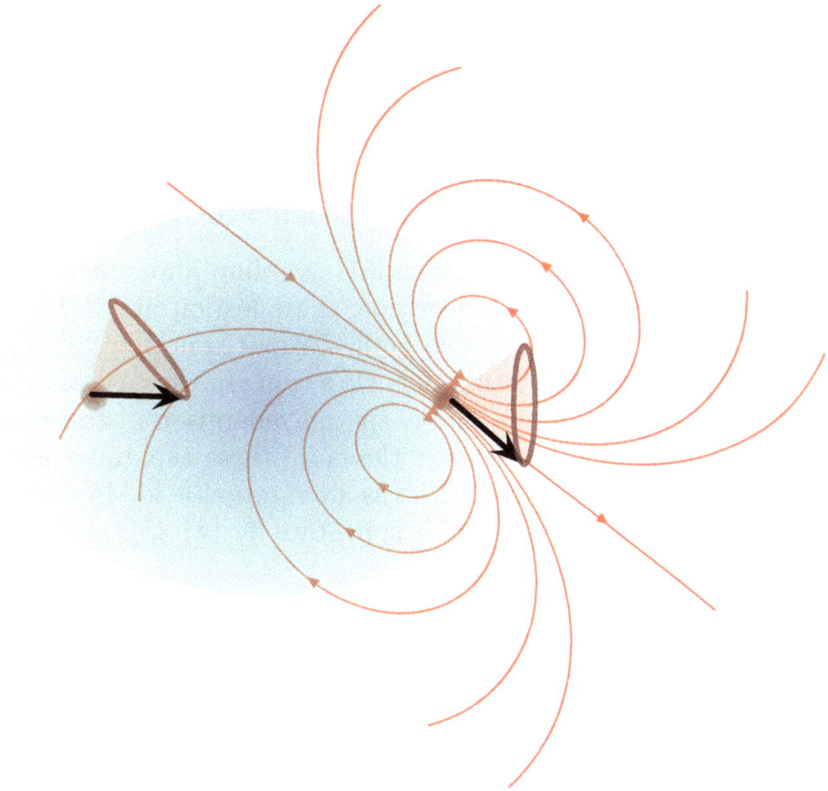

FIGURE 1.3.1. An illustration of two protons' Larmor spin-precession in a hydrogen molecule. Each proton perceives the other proton's magnetic field (directed red curves) as an externally applied magnetic field, and Larmor spin-precesses (cones with arrow) around the external magnetic field line.

Is there a principal difference between the magnetic field applied by an external apparatus and the magnetic field applied by an external particle? Taking the example of a hydrogen molecule, the magnetic field applied onto the proton by an other particle 74 pm away is principally not different from the magnetic field applied by an external apparatus. Taking the example of Knight shift measurements, the magnetic field applied onto the proton by a macroscopically delocalized electron wave is principally not different from the magnetic field applied by an external apparatus. It is clear from these examples that the magnetic field of one particle induces a Larmor spin-precession of the other particle; i.e. each particle of the involved ISC pair undergoes Larmor spin-precession. This concept is illustrated in figure 1.3.1 for a hydrogen molecule's two protons: the two proton spins are aligned on the average, while their actual magnetic moment is in a precession. One may observe in figure 1.3.1 an interesting coupling dynamics for the x component of the spin, when the x axis connects the two protons. Taking the magnetic moment vector of the left proton to point along towards the right proton, the right proton perceives a magnetic field pointing along the x axis, and its magnetic moment vector will thus be in a precession around the x axis. In turn, the left proton perceives a magnetic field coming at an angle with respect to the x axis, causing a tilted precession cone that rotates in sync around the x axis while the actual magnetic moment vector remains aligned with the x axis. In essence, the entire magnetic field configuration depicted in figure 1.3.1 is rotating around the x axis. A similar coupling dynamics can be worked out for the y and z spin directions.

For the purpose of our analysis, we do not need to know details of spin precession dynamics because in the following we shall work with the time-averaged spin measurements of an ISC pair.

We established through the above examples that the isotropic spin-correlation of separated particles involves Larmor spin-precession. At the same time, each involved particle has a principally measurable spin state, and thus we can investigate the origin of strictly N=2 coupling by taking into account the fact of individually measurable spin state.

1.4. Isotropic spin-coupling limit for incoherent electron states

Considering the phenomenological pair-wise ISC coupling limit, regardless of the involved particles being on the same orbital or not, the question arises whether the two electrons sharing the same orbital might also be in Larmor spin-precession with respect to the magnetic field generated by the other particle's spin. In the case of an antibonding molecular orbital, its two electrons have a large spatial separation distance because their wavefunction overlap region is mostly empty; therefore these separated electrons again perceive the magnetic field of the other electron as an external field. In turn, this implies Larmor precession and individually measurable spin state for two ISC electrons occupying an antibonding molecular orbital.

Let us finally consider an incoherent electron pair sharing the same bonding orbital; their wavefunctions overlap, and Zeeman split measurements yield the zero sum of the two oppositely oriented spins. Nevertheless, it may be possible to individually measure each electron's spin state by firstly separating them in such a way that does not disrupt their spin state, and then measuring each electron's spin state. The principal feasibility of such individual electron spin state measurement is demonstrated by reference **[7]**, whose authors studied the molecular photo-dissociation of H_2 and D_2 under linearly polarized incident light, employing 33.66 eV photon energy. They measured the angular correlation function of electromagnetic Lyman-alpha radiation produced by the resulting atom pair in order to find out whether the atom pair is entangled or not. The authors of **[7]** conclude that an entangled electron pair is produced through the photo-dissociation of a hydrogen molecule, and this entanglement originates from their molecular state. The results of **[7]** thus demonstrate that it is principally possible to photo-dissociate a bonding orbital occupying electron pair, without breaking their entanglement, and then measure their individual spin state.

Up to now, we established that electrons occupying incoherent states can be always treated as electrons with individually measurable spin state. In the following, we consider the implications of individual spin measurability, based on the methodology of reference **[4]**. By definition, N electrons are said to be isotropically spin-correlated (ISC), if a measurement made in an *arbitrary* direction on *one* of the particles allows us to predict with certainty the spin value of each of the other $N-1$ particles for the same direction.

THEOREM 1. *Incoherent ISC states exist only for $N = 2$.*

Essentially, to show that ISC states exist only for $N = 2$, it is sufficient to prove that it is impossible to have three such particles. The impossibility of three ISC particles also excludes the possibility of $N \geq 3$ ISC particles.

Suppose that an ISC state exists for $N = 3$. We demonstrate in the following paragraphs that this assumption leads to a mathematical contradiction.

In the interest of clarity, assume without loss of generality that the three ISC particles are such that they are detected to be in $(+,+,+)$ correlation for an arbitrary measurement direction. Later we will generalize the proof to any other correlation type. Define the x axis along this arbitrary direction, and define the z axis in any orthogonal direction to x. We will perform further spin measurements in the $x-z$ plane. Spin measurements in orthogonal directions are statistically independent. Although we know a given particle spin to be $|+\rangle$ along the x axis, a subsequent spin measurement along the z axis of the apparatus gives $\frac{1}{2}$ probability of measuring $|-\rangle$ state. In general, a spin state in direction 2θ with respect to the x axis, given that it is in the state $|+\rangle$ with respect to the x axis, can be constructed from the rotation R and is given by $R|+\rangle = \cos\theta\,|+\rangle - \sin\theta\,|-\rangle$. Therefore, in direction 2θ the probability of measuring $|+\rangle$ state is $\cos^2\theta$ and of measuring $|-\rangle$ is $\sin^2\theta$. Taking the $(x, 2\theta)$ direction with respect to two spin correlated particles, the joint probabilities are $P(+,+) = \frac{1}{2}\cos^2\theta$ and $P(+,-) = \frac{1}{2}\sin^2\theta$. Similarly, for the ket $|-\rangle$, $R|-\rangle = \sin\theta\,|+\rangle + \cos\theta\,|-\rangle$ and the joint probabilities are $P(-,-) = \frac{1}{2}\cos^2\theta$ and $P(-,+) = \frac{1}{2}\sin^2\theta$. In principle, if three ISC particles exist, a sequence of spin correlated measurements in the directions $2\theta_1$, $2\theta_2$, $2\theta_3$ can be performed on the three entangled particles. Let $(s_1(\theta_1), s_2(\theta_2), s_3(\theta_3))$ represent each particle's observed spin values in the three different directions. Recall that the above stated spin correlation implies that if any particle is measured to be in the $s_i(\theta_i) = |+\rangle$ spin state, the probability of measuring an other particle in the $s_j(\theta_j) = |-\rangle$ spin state becomes $\frac{1}{2}\sin^2(\theta_j - \theta_i)$.

Given that $s_n(\theta_n) = |\pm\rangle$ for each n, there exists only two possible values for each measurement, which we associate with "spin-up" and "spin-down" respectively. Hence, for three measurements there are a total of 8 possibilities. In particular,

$$\{(+,+,-),(+,-,-)\} \subset \{(+,+,-),(+,-,-),(-,+,-),(+,-,+)\}$$

implies the following probability relationship:

$$P\{(+,+,-),(+,-,-)\} \leq P\{(+,+,-),(+,-,-),(-,+,-),(+,-,+)\}$$

Consider the meaning of various subsets in the above inequality:

- The $\{(+,+,-),(+,-,-)\}$ subset is interpreted as follows: we measured the spin of particle 1 to be in $|+\rangle$ state and particle 3 to be in $|-\rangle$ state. The corresponding probability is $\frac{1}{2}\sin^2(\theta_3 - \theta_1)$.
- The $\{(+,+,-),(-,+,-)\}$ subset is interpreted as follows: we measured the spin of particle 2 to be in $|+\rangle$ state and particle 3 to be in $|-\rangle$ state. The corresponding probability is $\frac{1}{2}\sin^2(\theta_3 - \theta_2)$.
- The $\{(+,-,-),(+,-,+)\}$ subset is interpreted as follows: we measured the spin of particle 1 to be in $|+\rangle$ state and particle 2 to be in $|-\rangle$ state. The corresponding probability is $\frac{1}{2}\sin^2(\theta_2 - \theta_1)$.

Substituting the above terms into the above inequality, we arrive at

$$\frac{1}{2}\sin^2(\theta_3 - \theta_1) \leq \frac{1}{2}\sin^2(\theta_3 - \theta_2) + \frac{1}{2}\sin^2(\theta_2 - \theta_1)$$

Taking $\theta_3 - \theta_2 = \theta_2 - \theta_1 = \frac{\pi}{6}$ and $\theta_3 - \theta_1 = \frac{\pi}{3}$, the above inequality gives $\frac{1}{2} \geq \frac{3}{4}$, which is a contradiction. Therefore, three particles cannot all be in the same spin state with probability 1.

We note that **the specific value of electron spin plays no role in the proof of theorem 1, which indicates the irrelevance of Pauli's spin value based classification.**

The real reason behind the exclusion principle is the isotropic spin correlation requirement of eigenstate-occupying electrons, along with the impossibility of more than two such electrons in the case of individually measurable spins.

Remark: The proof of the above theorem was worked out for $(+,+,+)$ or $(-,-,-)$ type spin correlation. To generalize the proof, suppose that the ISC particles are measured to be $(+,-,+)$ along an arbitrary measurement direction. Then the spin outcomes in the three different directions θ_1, θ_2, θ_3 can be written as:

$$\{(+,-,-),(+,+,-)\} \subset \{(+,-,-),(+,+,-),(-,-,-),(+,+,+)\}$$

Essentially, this means that we flipped the $|+\rangle$ to $|-\rangle$ to represent the state of particle number two. Applying the same probability argument as before, but noting that $P\{(+,-,-),(-,-,-)\} = \frac{1}{2}\cos^2(\theta_3-\theta_2)$, the inequality becomes

$$\frac{1}{2}\sin^2(\theta_3-\theta_1) \leq \frac{1}{2}\cos^2(\theta_3-\theta_2) + \frac{1}{2}\cos^2(\theta_2-\theta_1)$$

Then upon taking $\theta_3-\theta_2 = \theta_2-\theta_1 = \frac{\pi}{2}-\frac{\pi}{6}$ and $\theta_3-\theta_1 = \pi-\frac{\pi}{3}$ gives as before $\frac{1}{2} \geq \frac{3}{4}$, which is a contradiction.

1.5. The isotropy of anti-parallel and parallel spin correlations

We already mentioned the two known types of isotropic spin-correlations: the anti-parallel correlated "spin singlet" state and the parallel correlated "spin triplet" state. The above discussed coupling limit theorem applies to both types. Here, we show that both parallel and anti-parallel correlations are isotropic. Understanding parallel spin correlations is the key to understanding electrons' coherent state establishment, which proceeds from this correlation type.

Usually, we identify a spin state measurement outcome by the following notation:

$$|+\rangle = \begin{pmatrix} 1 \\ 0 \end{pmatrix} \text{ and } |-\rangle = \begin{pmatrix} 0 \\ 1 \end{pmatrix}$$

But what do we really mean by "spin up" and "spin down"? Consider the z-axis spin angular momentum operator S_z defined by

$$S_z|+\rangle_z = \frac{1}{2}|+\rangle_z \text{ and } S_z|-\rangle_z = -\frac{1}{2}|-\rangle_z \ .$$

where the $\frac{1}{2}$ factor is the spin eigenvalue. More explicitly, we can write this spin operator in a matrix form:

$$S_z|+\rangle_z = \begin{pmatrix} \frac{1}{2} & 0 \\ 0 & -\frac{1}{2} \end{pmatrix}\begin{pmatrix} 1 \\ 0 \end{pmatrix} = \frac{1}{2}\begin{pmatrix} 1 \\ 0 \end{pmatrix} \text{ and } S_z|-\rangle_z = \begin{pmatrix} \frac{1}{2} & 0 \\ 0 & -\frac{1}{2} \end{pmatrix}\begin{pmatrix} 0 \\ 1 \end{pmatrix} = -\frac{1}{2}\begin{pmatrix} 0 \\ 1 \end{pmatrix}$$

and consequently

$$(1.5.1)\quad S_z|+\rangle_z = \begin{pmatrix} \frac{1}{2} & 0 \\ 0 & -\frac{1}{2} \end{pmatrix}\begin{pmatrix} 1 \\ 0 \end{pmatrix} = \begin{pmatrix} -\frac{1}{2} & 0 \\ 0 & \frac{1}{2} \end{pmatrix}\begin{pmatrix} -1 \\ 0 \end{pmatrix} = \frac{1}{2}\begin{pmatrix} 1 \\ 0 \end{pmatrix} = -\frac{1}{2}\begin{pmatrix} -1 \\ 0 \end{pmatrix}$$

From this perspective, the S_z versus $-S_z$ operators represent spin measurements along the z and $-z$ directions, and the associated unit eigenvectors are $\begin{pmatrix} 1 \\ 0 \end{pmatrix}$ versus $\begin{pmatrix} -1 \\ 0 \end{pmatrix}$. Using analogous formalism, we can define x-axis and y-axis spin angular momentum operators as well: these are listed in table 2.

The usefulness of this quantum mechanical representation is seen when we calculate spin state probabilities along various directions. The electron spin can be measured along

TABLE 2. Spin operators and eigenvectors

Spin operator	Eigenvectors	Eigenvalues
$S_x = \begin{pmatrix} 0 & \frac{1}{2} \\ \frac{1}{2} & 0 \end{pmatrix}$	$\begin{pmatrix} \frac{1}{\sqrt{2}} \\ \frac{1}{\sqrt{2}} \end{pmatrix}, \begin{pmatrix} \frac{-1}{\sqrt{2}} \\ \frac{1}{\sqrt{2}} \end{pmatrix}$	$\frac{1}{2}, -\frac{1}{2}$
$S_y = \begin{pmatrix} 0 & \frac{-i}{2} \\ \frac{i}{2} & 0 \end{pmatrix}$	$\begin{pmatrix} \frac{1}{\sqrt{2}} \\ \frac{i}{\sqrt{2}} \end{pmatrix}, \begin{pmatrix} \frac{i}{\sqrt{2}} \\ \frac{1}{\sqrt{2}} \end{pmatrix}$	$\frac{1}{2}, -\frac{1}{2}$
$S_z = \begin{pmatrix} \frac{1}{2} & 0 \\ 0 & \frac{-1}{2} \end{pmatrix}$	$\begin{pmatrix} 1 \\ 0 \end{pmatrix}, \begin{pmatrix} 0 \\ 1 \end{pmatrix}$	$\frac{1}{2}, -\frac{1}{2}$

any direction, and orthogonal measurements give independent results. Firstly, we measure the electron spin state along the z direction. Before the measurement, a general spin state is written as

$$|s\rangle = a\,|+\rangle_z + b\,|-\rangle_z = a\begin{pmatrix} 1 \\ 0 \end{pmatrix} + b\begin{pmatrix} 0 \\ 1 \end{pmatrix}$$

Without any measurement or filtering, we have $a = b = \frac{1}{\sqrt{2}}$, which means $\frac{1}{2}$ probability for each spin eigenstate. Acting with the operator S_z on $|s\rangle$ means observing the average spin state along the z axis:

$$S_z\,|s\rangle = \frac{a}{2}\begin{pmatrix} 1 \\ 0 \end{pmatrix} - \frac{b}{2}\begin{pmatrix} 0 \\ 1 \end{pmatrix}$$

Suppose that we want to know just the coefficient b of the spin state. This is obtained by the following expression:

$$_z\langle -|\, s\rangle = a\begin{pmatrix} 0 & 1 \end{pmatrix}^{\dagger}\begin{pmatrix} 1 \\ 0 \end{pmatrix} + b\begin{pmatrix} 0 & 1 \end{pmatrix}^{\dagger}\begin{pmatrix} 0 \\ 1 \end{pmatrix} = b$$

The above method can be checked for all other directions, and shows that our eigenvectors are orthogonal for any measurement direction.

Suppose that we also want to know the spin along the x direction. We write the above eigenvectors as follows:

$$|+\rangle_z = \begin{pmatrix} 1 \\ 0 \end{pmatrix} = \frac{1}{\sqrt{2}}\left[\begin{pmatrix} \frac{1}{\sqrt{2}} \\ \frac{1}{\sqrt{2}} \end{pmatrix} - \begin{pmatrix} \frac{-1}{\sqrt{2}} \\ \frac{1}{\sqrt{2}} \end{pmatrix}\right] = \frac{1}{\sqrt{2}}\left[|+\rangle_x - |-\rangle_x\right]$$

$$|-\rangle_z = \begin{pmatrix} 0 \\ 1 \end{pmatrix} = \frac{1}{\sqrt{2}}\left[\begin{pmatrix} \frac{1}{\sqrt{2}} \\ \frac{1}{\sqrt{2}} \end{pmatrix} + \begin{pmatrix} \frac{-1}{\sqrt{2}} \\ \frac{1}{\sqrt{2}} \end{pmatrix}\right] = \frac{1}{\sqrt{2}}\left[|+\rangle_x + |-\rangle_x\right]$$

The above result means that if we filter the electron to be in the $|+\rangle_z$ state, a subsequent spin measurement along the x axis yields equal probability of finding the electron in any of the two spin states. Similarly, upon filtering the electron to be in the $|-\rangle_z$ state, a subsequent spin measurement along the x axis yields again equal probability of finding the electron in any of the two spin states. Therefore, our mathematical notation captures the independence of spin measurements in orthogonal directions. In the $x-z$ plane, a rotation of spin measurement direction is described by a rotation of eigenvectors according to the $R = \begin{pmatrix} \cos\theta & -\sin\theta \\ \sin\theta & \cos\theta \end{pmatrix}$ rotation matrix.

In the $y-z$ plane, a rotation of spin measurement direction is described by a rotation of eigenvectors between real and imaginary axes. This rotation is described in the appendix.

Having clarified the spin state notation for one electron, let us move on to two-electron systems. The quantum mechanical state of a two-electron system is represented by the tensor product of each electron state. Therefore, the two-electron spin state is represented by a tensor product of individual electron spin states.

DEFINITION. The spin-singlet state of two electrons is defined as the anti-parallel orientation of their spin states:

$$|\uparrow\downarrow\rangle \equiv \frac{1}{\sqrt{2}}\left[\begin{pmatrix}1\\0\end{pmatrix}\otimes\begin{pmatrix}0\\1\end{pmatrix}-\begin{pmatrix}0\\1\end{pmatrix}\otimes\begin{pmatrix}1\\0\end{pmatrix}\right] \tag{1.5.2}$$

where we formed the two-electron spin state via an anti-symmetric combination of anti-parallel oriented spin states. The advantage of this definition is that it not only captures the anti-parallel pairing process, but also implicitly captures the isotropy associated with the pairing. In the following, we prove this isotropy.

THEOREM 2. *A spin-singlet state defined according to equation 1.5.2 means that the two electrons' spin angular momenta are isotropically correlated and the resulting two-electron spin state is rotationally invariant.*

For a spin measurement in the z direction, equation 1.5.2 remains in its above-defined form:

$$|\uparrow\downarrow\rangle_z = \frac{1}{\sqrt{2}}\left[\begin{pmatrix}1\\0\end{pmatrix}\otimes\begin{pmatrix}0\\1\end{pmatrix}-\begin{pmatrix}0\\1\end{pmatrix}\otimes\begin{pmatrix}1\\0\end{pmatrix}\right]$$

For a spin measurement in the x direction, equation 1.5.2 is written as:

$$|\uparrow\downarrow\rangle_x = \frac{1}{\sqrt{2}}\left[\begin{pmatrix}\frac{1}{\sqrt{2}}\\\frac{1}{\sqrt{2}}\end{pmatrix}\otimes\begin{pmatrix}\frac{-1}{\sqrt{2}}\\\frac{1}{\sqrt{2}}\end{pmatrix}-\begin{pmatrix}\frac{-1}{\sqrt{2}}\\\frac{1}{\sqrt{2}}\end{pmatrix}\otimes\begin{pmatrix}\frac{1}{\sqrt{2}}\\\frac{1}{\sqrt{2}}\end{pmatrix}\right]=$$
$$=\frac{1}{\sqrt{2}}\left(\frac{1}{2}+\frac{1}{2}\right)\left[\begin{pmatrix}1\\0\end{pmatrix}\otimes\begin{pmatrix}0\\1\end{pmatrix}-\begin{pmatrix}0\\1\end{pmatrix}\otimes\begin{pmatrix}1\\0\end{pmatrix}\right]$$

To see the above result, one may expand $\begin{pmatrix}\frac{1}{\sqrt{2}}\\\frac{1}{\sqrt{2}}\end{pmatrix}\otimes\begin{pmatrix}\frac{-1}{\sqrt{2}}\\\frac{1}{\sqrt{2}}\end{pmatrix}$ as:

$$\frac{1}{2}\begin{pmatrix}1\\0\end{pmatrix}\begin{pmatrix}0\\1\end{pmatrix}-\frac{1}{2}\begin{pmatrix}0\\1\end{pmatrix}\begin{pmatrix}1\\0\end{pmatrix}-\frac{1}{2}\begin{pmatrix}1\\0\end{pmatrix}\begin{pmatrix}1\\0\end{pmatrix}+\frac{1}{2}\begin{pmatrix}0\\1\end{pmatrix}\begin{pmatrix}0\\1\end{pmatrix}$$

Similarly, one may expand $-\begin{pmatrix}\frac{-1}{\sqrt{2}}\\\frac{1}{\sqrt{2}}\end{pmatrix}\otimes\begin{pmatrix}\frac{1}{\sqrt{2}}\\\frac{1}{\sqrt{2}}\end{pmatrix}$ as:

$$\frac{1}{2}\begin{pmatrix}1\\0\end{pmatrix}\begin{pmatrix}0\\1\end{pmatrix}-\frac{1}{2}\begin{pmatrix}0\\1\end{pmatrix}\begin{pmatrix}1\\0\end{pmatrix}+\frac{1}{2}\begin{pmatrix}1\\0\end{pmatrix}\begin{pmatrix}1\\0\end{pmatrix}-\frac{1}{2}\begin{pmatrix}0\\1\end{pmatrix}\begin{pmatrix}0\\1\end{pmatrix}$$

The equivalence between $|\uparrow\downarrow\rangle_z$ and $|\uparrow\downarrow\rangle_x$ follows by adding the above expressions. We thus get exactly the same result for the spin measurement in the z and x directions. Regarding an arbitrary spin measurement direction in the $x-z$ plane, the eigenvectors transform via the above explained R rotation matrix, and the reader may validate that $\frac{1}{\sqrt{2}}\left[R\begin{pmatrix}1\\0\end{pmatrix}\otimes R\begin{pmatrix}0\\1\end{pmatrix}-R\begin{pmatrix}0\\1\end{pmatrix}\otimes R\begin{pmatrix}1\\0\end{pmatrix}\right]$ evaluates to equation 1.5.2. The same result can be worked out also for the y direction: this calculation is shown in the appendix. Therefore, equation 1.5.2 expresses isotropic spin entanglement in all directions. The proof of theorem 2 is complete.

We note that spin-singlet states are also invariant under the $SL(2,\mathbb{C})$ group action, which is a more general symmetry than just rotational invariance. This symmetry is studied in reference **[5]**.

DEFINITION. The spin-triplet state of two electrons is defined as the parallel orientation of their spin states:

$$|\uparrow\uparrow\rangle \equiv \frac{1}{\sqrt{2}}\left[\begin{pmatrix}1\\0\end{pmatrix}\otimes\begin{pmatrix}1\\0\end{pmatrix}+\begin{pmatrix}0\\1\end{pmatrix}\otimes\begin{pmatrix}0\\1\end{pmatrix}\right] \tag{1.5.3}$$

where we formed the two-electron spin state via a combination of parallel oriented spin states. Again, the advantage of this definition is that it not only captures the parallel pairing process, but also implicitly captures the isotropy associated with the pairing.

THEOREM 3. *A spin-triplet state defined according to equation 1.5.3 means that the two electrons' spin angular momenta are isotropically correlated and the resulting two-electron spin state is rotationally invariant.*

For a spin measurement in the z direction, equation 1.5.3 remains in its above-defined form:

$$|\uparrow\uparrow\rangle_z = \frac{1}{\sqrt{2}}\left[\begin{pmatrix}1\\0\end{pmatrix}\otimes\begin{pmatrix}1\\0\end{pmatrix}+\begin{pmatrix}0\\1\end{pmatrix}\otimes\begin{pmatrix}0\\1\end{pmatrix}\right]$$

For a spin measurement in the x direction, equation 1.5.3 is written as:

$$|\uparrow\uparrow\rangle_x = \frac{1}{\sqrt{2}}\left[\begin{pmatrix}\frac{1}{\sqrt{2}}\\\frac{1}{\sqrt{2}}\end{pmatrix}\otimes\begin{pmatrix}\frac{1}{\sqrt{2}}\\\frac{1}{\sqrt{2}}\end{pmatrix}+\begin{pmatrix}\frac{-1}{\sqrt{2}}\\\frac{1}{\sqrt{2}}\end{pmatrix}\otimes\begin{pmatrix}\frac{-1}{\sqrt{2}}\\\frac{1}{\sqrt{2}}\end{pmatrix}\right]=$$

$$=\frac{1}{\sqrt{2}}\left(\frac{1}{2}+\frac{1}{2}\right)\left[\begin{pmatrix}1\\0\end{pmatrix}\begin{pmatrix}1\\0\end{pmatrix}+\begin{pmatrix}0\\1\end{pmatrix}\begin{pmatrix}0\\1\end{pmatrix}\right]$$

To see the above result, one may expand $\begin{pmatrix}\frac{1}{\sqrt{2}}\\\frac{1}{\sqrt{2}}\end{pmatrix}\otimes\begin{pmatrix}\frac{1}{\sqrt{2}}\\\frac{1}{\sqrt{2}}\end{pmatrix}$ as:

$$\frac{1}{2}\begin{pmatrix}1\\0\end{pmatrix}\begin{pmatrix}0\\1\end{pmatrix}+\frac{1}{2}\begin{pmatrix}0\\1\end{pmatrix}\begin{pmatrix}1\\0\end{pmatrix}+\frac{1}{2}\begin{pmatrix}1\\0\end{pmatrix}\begin{pmatrix}1\\0\end{pmatrix}+\frac{1}{2}\begin{pmatrix}0\\1\end{pmatrix}\begin{pmatrix}0\\1\end{pmatrix}$$

Similarly, one may expand $\begin{pmatrix}\frac{-1}{\sqrt{2}}\\\frac{1}{\sqrt{2}}\end{pmatrix}\otimes\begin{pmatrix}\frac{-1}{\sqrt{2}}\\\frac{1}{\sqrt{2}}\end{pmatrix}$ as:

$$-\frac{1}{2}\begin{pmatrix}1\\0\end{pmatrix}\begin{pmatrix}0\\1\end{pmatrix}-\frac{1}{2}\begin{pmatrix}0\\1\end{pmatrix}\begin{pmatrix}1\\0\end{pmatrix}+\frac{1}{2}\begin{pmatrix}1\\0\end{pmatrix}\begin{pmatrix}1\\0\end{pmatrix}+\frac{1}{2}\begin{pmatrix}0\\1\end{pmatrix}\begin{pmatrix}0\\1\end{pmatrix}$$

The equivalence between $|\uparrow\uparrow\rangle_z$ and $|\uparrow\uparrow\rangle_x$ follows by adding the above expressions. We thus get exactly the same result for the spin measurement in the z and x directions. Regarding an arbitrary spin measurement direction in the $x-z$ plane, the eigenvectors transform via the R rotation matrix, and the reader may validate that $\frac{1}{\sqrt{2}}\left[R\begin{pmatrix}1\\0\end{pmatrix}\otimes R\begin{pmatrix}1\\0\end{pmatrix}+R\begin{pmatrix}0\\1\end{pmatrix}\otimes R\begin{pmatrix}0\\1\end{pmatrix}\right]$ evaluates to equation 1.5.3. The analogous calculation for the y direction is also shown in the appendix. Therefore, equation 1.5.3 expresses isotropic spin entanglement in all directions. The proof of theorem 3 is complete.

In summary, the usual formalism of quantum mechanics demonstrates that both spin-singlet and spin-triplet states are rotationally invariant. Therefore, both states lead to a well-defined electron energy eigenstate, which is the pre-requisite for pair-wise electron occupancy of a quantum mechanical orbital. In the past, only the spin-singlet state was considered, due to malformed electron statistics rules. In contrast, our results show that an electron pair of any orbital may be found in spin-triplet state as well. We emphasize that the parallel spin correlation of a spin-triplet state is still an averaged value: the two electrons are in incoherent state, involving Larmor precession. However, this constellation of parallel averaged spins is precursor to the establishment of strictly parallel spin orientation, which is the subject of the following section.

1.6. Spin statistics for coherent electron states

1.6.1. Methodology. Our goal is to determine whether N electrons could indeed form a coherent wavefunction of the form $\psi = \sqrt{n}e^{i\varphi}|s\rangle$, where φ is the quantum mechanical phase of the common wavefunction, $n \equiv \frac{N}{V}$ is its electron density, and $|s\rangle$ is its spin state. Up to now, this question was approached dogmatically via Pauli's spin value based classification scheme, but we already showed that the electron spin value is irrelevant to the Pauli exclusion principle.

Directly employing the formalism of quantum mechanics, it remains impossible to decide whether a coherent electron wavefunction would form. Using the Schrödinger equation, it is easy to see that having four electrons in the $1s$ orbital is lower energy configuration than having two electrons in the $1s$ and two electrons in the $2s$ orbitals. However, the generally observed beryllium configuration comprises two electrons in the $1s$ and two electrons in the $2s$ orbitals. Clearly, relying on Schrödinger equation based energy minimization is does not give the correct answer. What we are missing is the thermodynamic principle that determines whether coherent or incoherent electron state is preferred under a give set of conditions. One could try to remedy the situation by comparing the von Neumann entropy of coherent and incoherent electron states. Unfortunately, such an approach does not capture the entropy associated with the presence versus absence of individual electrons' Larmor precessions.

One possible approach is phenomenological. The analysis of AC Josephson effect in chapter 4 unambigously demonstrates the existence of N electrons' coherent wavefunction. Readers preferring the phenomenological approach may check the proofs in chapter 4 and skip to the next section.

The other possible approach is to look beyond the operator-based formalism of quantum mechanics, which deals with measurement outcomes on wavefunctions, and to consider the wavefunction generating process. In other words, a brief discussion of wave-particle duality is required in order to understand electron-electron interaction at the microscopic scale. Readers who are interested in such electron-electron interaction dynamics, which determines whether a coherent state is possible, should continue reading this section.

1.6.2. A review of the Darwin Lagrangian. In this section we shall make use the so-called Darwin Lagrangian, which we therefore briefly review. The Darwin Lagrangian is well-known for modeling the interaction among a large number of massive charged particles. It is defined as follows:

$$\mathscr{L}_D = \sum_{a=1}^{N} \left\{ \frac{1}{2} m_a v_a^2 + \frac{1}{2} \left[e\boldsymbol{A}_a \left(\boldsymbol{r}_a \right) \cdot \boldsymbol{v}_a - e_a \phi_a \left(\boldsymbol{r}_a \right) \right] \right\} \tag{1.6.1}$$

where $\boldsymbol{r}_a$ is the vectorial position of a given particle, e_a is its charge value, $\boldsymbol{v}_a$ is its velocity, and m_a is its mass. $\boldsymbol{A}_a \left(\boldsymbol{r}_a \right)$ and $\phi \left(\boldsymbol{r}_a \right)$ are the vector potential and Coulomb potential at position $\boldsymbol{r}_a$, and N is the total number of the interacting particles.

The Coulomb potential $\phi_a \left(\boldsymbol{r}_a \right)$ is given by:

$$\phi_a \left(\boldsymbol{r}_a \right) = \sum_{b \neq a}^{N} \frac{e_b}{r_{ab}}$$

where $r_{ab} = |\boldsymbol{r}_{ab}| = |\boldsymbol{r}_b - \boldsymbol{r}_a|$ is the euclidean distance from other particles.

Particles a and b moving parallel to their distance vector have no magnetic interaction, and consequently only the $\boldsymbol{v}_{b\perp}$ component of the charge velocity vector shall contribute to the vector potential at $\boldsymbol{r}_a$:

$$\boldsymbol{v}_{b\perp} = \boldsymbol{v}_b - (\boldsymbol{v}_b \cdot \boldsymbol{r}_{uab})\, \boldsymbol{r}_{uab}$$

where $\boldsymbol{v}_b$ is the charge velocity vector, $\boldsymbol{v}_{b\perp}$ is its orthogonal component to the distance vector $\boldsymbol{r}_{ab}$, and $\boldsymbol{r}_{uab} = \frac{\boldsymbol{r}_{ab}}{r_{ab}}$ is the unit length vector.

The $\boldsymbol{A}_{ab}(\boldsymbol{r}_a)$ vector potential contribution of particle b is therefore given by:

$$\boldsymbol{A}_{ab}(\boldsymbol{r}_a) = \frac{e_b \boldsymbol{v}_{b\perp}}{r_{ab}} =$$

$$= \frac{e_b \left[\boldsymbol{v}_b - (\boldsymbol{v}_b \cdot \boldsymbol{r}_{uab})\, \boldsymbol{r}_{uab}\right]}{r_{ab}}$$

We note that $\boldsymbol{A}_{ba} \cdot \boldsymbol{v}_b = \boldsymbol{A}_{ab} \cdot \boldsymbol{v}_a$.

In the above equations$\boldsymbol{A}_{ab}$ is the contribution of particle b to the vector potential $\boldsymbol{A}_a$. The overall vector potential $\boldsymbol{A}_a$ at $\mathbf{r}_a$ is the sum of contributions by all other particles:

$$\boldsymbol{A}_a(\boldsymbol{r}_a) = \sum_{b \neq a}^{N} \boldsymbol{A}_{ab}(\boldsymbol{r}_a) = \sum_{b \neq a}^{N} \frac{e_b \left[\boldsymbol{v}_b - (\boldsymbol{v}_b \cdot \boldsymbol{r}_{uab})\, \boldsymbol{r}_{uab}\right]}{r_{ab}} \tag{1.6.2}$$

The $\boldsymbol{A}_a$ and ϕ_a terms are halved in the Darwin Lagrangian in order to avoid a double counting of contributions to the collective interaction potential, considering that $\boldsymbol{A}_{ab} \cdot \boldsymbol{v}_a = \boldsymbol{A}_{ba} \cdot \boldsymbol{v}_b$ and $\phi_{ab} = \phi_{ba}$.

1.6.3. A brief review of electron Zitterbewegung. The existence of electron Zitterbewegung was first suggested by De Broglie, who proposed the mc^2/h oscillation frequency named after him, and then directly described as a light-speed oscillation by Schrödinger. Reference [**10**] presents an experimentally validated Zitterbewegung model of the electron structure. As shown in [**10**], the electron spin is generated by its circular Zitterbewegung oscillation. This idea of the electron spin being generated by circular Zitterbewegung oscillation has a long history; reference [**14**] presents a thorough discussion of this topic. The Thomson scattering phenomenon is electron-light interaction in the low photon frequency limit: it measures the electron's "reduced Compton radius" size, which corresponds to the radius of light-speed charge circulation at the mc^2/h frequency.

In quantum theory, Zitterbewegung appears in the free electron a solution of the Dirac equation. A comprehensive introduction to the Dirac equation can be found in chapter 5 of reference [**10**]. Taking the Heisenberg picture, and solving the Dirac equation for the electron position yields a high frequency oscillation term, with the corresponding oscillation velocity being the speed of light. This constant velocity implies a circular Zitterbewegung oscillation at light-speed. The details of this calculation can be found for example on pages 322-323 of [**8**]. In modern quantum theory, Zitterbewegung oscillation shows up in the Gordon decomposition of the Dirac current, which splits the charge current into a part that arises from the motion of the center of mass, and a part that arises from gradients of the spin density. The equivalence between the Heisenberg picture based electron position calculation and the Gordon decomposition is discussed in reference [].

We now demonstrate that the electron's quantum mechanical wavelength is in fact the Lorentz-transformed spatial component of its Zitterbewegung oscillation. Consider an electron moving at kinetic speed v. In relation to light-speed, its speed is characterized by $\beta = \frac{v}{c}$, $\gamma_L = (1-\beta^2)^{-\frac{1}{2}}$ and rapidity w defined as $\gamma_L = \cosh w$. It follows that $\cosh^2 w - \sinh^2 w = 1$, $\tanh w = \beta$, and $\sinh w = \gamma_L \beta$.

In the electron's rest frame, its Zitterbewegung is a time-wise oscillation. A relativistic boost rotates the time and space axes into each other according to the following hyperbolic rotation matrix:

$$\begin{pmatrix} ct' \\ x' \end{pmatrix} = \begin{pmatrix} \cosh w & -\sinh w \\ -\sinh w & \cosh w \end{pmatrix} \begin{pmatrix} ct \\ x \end{pmatrix}$$

Therefore, the time-wise Zitterbewegung oscillation of the rest frame acquires a spatial oscillation component in the boosted reference frame. Specifically, the Zitterbewegung frequency of the rest frame is $\frac{\omega}{2\pi} = \frac{m_0 c^2}{h}$, and this is commonly referred to as the De Broglie frequency. The quantum mechanical wavenumber of the rest frame is: $k_0 = 0$. The corresponding wavenumber in the boosted frame is:

$$\frac{k}{2\pi} = \frac{\omega}{2\pi}\frac{\sinh w}{c} - k_0 \cosh w = \frac{\omega}{2\pi}\frac{\sinh w}{c}$$

Evaluating the right side of the above equation, we obtain:

$$\frac{k}{2\pi} = \frac{m_0 c^2}{h}\frac{\gamma_L v}{c^2}$$

Rearranging the above equation, we finally obtain:

$$\hbar k = (\gamma_L m_0)\, v = mv = p_{kinetic}$$

We recognize the above result as the basic postulate of quantum mechanics. However, it no longer needs to be a postulate: the appearing quantum mechanical wave is simply the Lorentz transformed component of the electron's Zitterbewegung oscillation. In this sense, all quantum mechanical wavelength measurements validate the Zitterbewegung structure of the electron, with the Zitterbewegung frequency being mc^2/h.

By definition, a coherent state of electrons means that they maintain the same quantum mechanical wavelength; this implies the coherence of their Zitterbewegung phases, and vice versa.

1.6.4. The stable equilibrium of coherent electron states. Considering the microscopic interaction among electrons, the Zitterbewegung Lagrangian of N interacting charges can be written analogously to the Darwin Lagrangian, but replacing the kinetic electron speed with the light-speed Zitterbewegung speed vector $\boldsymbol{c}$. Specifically, the Zitterbewegung Lagrangian must include the contributions of all other charges:

$$\mathscr{L}_N = -\sum_{a=1}^{N} \left[e_a \boldsymbol{A}_a \cdot \boldsymbol{c}_a - e_a V_a \right] = 0 \tag{1.6.3}$$

where $\mathscr{L}_N = 0$ is the condition for stable equilibrium. In the above equation, $\boldsymbol{A}$ is the vector potential generated by electrons' Zitterbewegung rotation at light speed, while V is the Coulomb potential at their charge surface. A detailed derivation of equation 1.6.3 can be found in chapter 4 of reference **[10]**.

At first sight, it might appear that we replaced the quantum mechanical approach by a classical approach, but that is not the case. Consider N coherent electrons forming a standing wave within a square potential well. We switch from the laboratory frame to a frame that is co-moving with the electrons; then each individual electron Zitterbewegung appears as illustrated in figure 1.6.1. Although we do not know the individual electron positions, their relative distance is defined by the enclosure. We can use equation 1.6.3 to examine the electric and magnetic interactions among electrons; the obtained equilibrium states must remain in equilibrium also from the laboratory frame perspective. Essentially, the proper use of Lorentz transformation allows us to work with the wavefunction-generating particle aspect, instead of performing a more difficult calculation directly on the electron wavefunction.

In the following, we use c=1 Natural units notation for convenience. Calling r_{ec} the electron charge radius, r_e the Zitterbewegung radius, m_e the electron mass, ω_e the Zitterbewegung angular speed and $\boldsymbol{c}$ the charge velocity vector in vacuum, we relate these values to each other in Natural units notation:

$$V = \frac{e}{r_{ec}} = \frac{e}{\alpha r_e} = \frac{1}{er_e}$$

$$\boldsymbol{A} = \frac{e\boldsymbol{c}}{\alpha r_e} = \frac{\boldsymbol{c}}{er_e}$$

where α is the electromagnetic fine structure constant.

The relativistic momentum of the electron charge is always given by $e\boldsymbol{A}_a = m_a\boldsymbol{c}_a$. Calling m_o the electron rest mass, the kinetic energy term of any given electron can be written as:

$$e\boldsymbol{A}_a \cdot \boldsymbol{c}_a = m_a\boldsymbol{c}_a \cdot \boldsymbol{c}_a = m_a = \gamma m_0 = \frac{m_0}{\sqrt{1-v_a^2}} \simeq m_0 + \frac{1}{2}m_0 v_a^2 \tag{1.6.4}$$

The above result means that the $e_a\boldsymbol{A}_a \cdot \boldsymbol{c}_a$ term already incorporates the electron kinetic energy. Equation 1.6.3 therefore gives the complete Lagrangian equation: unlike in the classical Darwin Lagrangian case, there is no additional $\frac{1}{2}m_a v_a^2$ term.

Using the $\mathscr{L}_N = 0$ is the condition for stable equilibrium, we shall demonstrate the existence of a coherent state of N electrons, as a stable equilibrium state.

THEOREM 4. *In the absence of noise, N coherent electrons form a stable equilibrium state*

When $N = 1$, the Lagrangian expression becomes $\mathscr{L}_a = -\left[e_a\boldsymbol{A}_a \cdot \boldsymbol{c}_a - e_a V_a\right]$. $\mathscr{L}_a$ is always zero as a consequence of the two Aharonov-Bohm equations that relate the Zitterbewegung rotation phase φ to the electromagnetic potentials. To see this, let us calculate the differential $d\varphi$ term, keeping in mind that the Zitterbewegung phase is equivalent to the quantum mechanical wavefunction's phase.

On the one hand, the $d\varphi$ phase change can be calculated from the magnetic Aharonov-Bohm equation:

$$d\varphi = e_a\boldsymbol{A}_a \cdot \boldsymbol{dl} = e_a\boldsymbol{A}_a \cdot \boldsymbol{c_a}dt$$

On the other hand, the same $d\varphi$ phase change can be calculated from the electric Aharonov-Bohm equation:

$$d\varphi = e_a V_a dt$$

Dividing the above two equations by dt, we can equate them. Consequently:

$$\mathscr{L}_a = -\left[e_a\boldsymbol{A}_a \cdot \boldsymbol{c}_a - e_a V_a\right] = 0 \tag{1.6.5}$$

$$e_a \boldsymbol{A}_a \cdot \boldsymbol{c}_a = e_a V_a$$

The above result means that each individual electron's total energy is the same as the potential energy on the surface of its electric charge.

We now evaluate $\boldsymbol{A}_a(\boldsymbol{r}_a)$ and $V_a(\boldsymbol{r}_a)$ for N interacting electrons:

$$\boldsymbol{A}_a(\boldsymbol{r}_a) = \frac{e_a \boldsymbol{c}_a}{\alpha r_{ea}} + \frac{1}{2}\sum_{i \neq j} \frac{e_b \boldsymbol{c}_{b\perp}}{r_{ab}} = \tag{1.6.6}$$

$$= \frac{e_a \boldsymbol{c}_a}{\alpha r_{ea}} + \frac{1}{2}\sum_{a \neq b} \frac{e_b \left[\boldsymbol{c}_b - (\boldsymbol{c}_b \cdot \boldsymbol{r}_{uab}) \boldsymbol{r}_{uab}\right]}{r_{ab}}$$

$$V_a(\boldsymbol{r}_a) = \frac{e_a}{\alpha r_{ea}} + \frac{1}{2}\sum_{a \neq b} \frac{e_b}{r_{ab}} \tag{1.6.7}$$

In both of these last two equations, the first terms are the self-interaction contributions. In equation 1.6.6, the $\boldsymbol{c}_{b\perp}$ component of the charge velocity vector $\boldsymbol{c}_b$ is orthogonal to the distance vector $\boldsymbol{r}_{ab}$. Analogously to the Darwin Lagrangian case, only this $\boldsymbol{c}_{b\perp}$ component will contribute to the value of the vector potential $\boldsymbol{A}_a(\boldsymbol{r}_a)$.

The $e_a \boldsymbol{A}_a \cdot \boldsymbol{c}_a$and $e_a V_a$ terms of equation 1.6.3 can now be calculated as follows:

$$e_a \boldsymbol{A}_a \cdot \boldsymbol{c}_a = \frac{e_a^2}{\alpha r_{ea}} \boldsymbol{c}_a^2 + \frac{1}{2}\sum_{a \neq b} \frac{e_a e_b \left[\boldsymbol{c}_b - (\boldsymbol{c}_b \cdot \boldsymbol{r}_{uab}) \boldsymbol{r}_{uab}\right] \cdot \boldsymbol{c}_a}{r_{ab}}$$

$$e_a V_a = \frac{e_a^2}{\alpha r_{ea}} + \frac{1}{2}\sum_{a \neq b} \frac{e_a e_b}{r_{ab}}$$

where $e_a^2 = \alpha$ and $\boldsymbol{c}_a^2 = 1$.

Since all charged particles are electrons, it is possible to write their charges as $e_a e_b = \alpha$, and the expression for $\mathscr{L}_N$ becomes:

$$\mathscr{L}_N = -\sum_{a=1}^{N} \left[\frac{e_a^2}{\alpha r_{ea}} \boldsymbol{c}_a^2 - \frac{e_a^2}{\alpha r_{ea}} + \frac{1}{2}\sum_{a \neq b} \frac{\alpha \left[\boldsymbol{c}_a \cdot \boldsymbol{c}_b - (\boldsymbol{c}_a \cdot \boldsymbol{r}_{uab})(\boldsymbol{c}_b \cdot \boldsymbol{r}_{uab}) - 1\right]}{r_{ab}}\right] \tag{1.6.8}$$

Considering that the charges speed is always $c = 1$, the first two terms of the above expression cancel out. Equation 1.6.8 therefore simplifies to:

$$\mathscr{L}_N = -\frac{1}{2}\sum_{a=1}^{N}\sum_{a \neq b} \frac{\alpha \left[\boldsymbol{c}_a \cdot \boldsymbol{c}_b - (\boldsymbol{c}_a \cdot \boldsymbol{r}_{uab})(\boldsymbol{c}_b \cdot \boldsymbol{r}_{uab}) - 1\right]}{r_{ab}}$$

This $\mathscr{L}_N$ Lagrangian is zero (i.e. minimized) for a collection of coherent electrons where the Zitterbewegung phase is the same for all electron charges, and their Zitterbewegung planes are parallel to each other. In this case $\boldsymbol{c}_a \cdot \boldsymbol{c}_b = 1$, and $\boldsymbol{c}_a \cdot \boldsymbol{r}_{uab} = \boldsymbol{c}_b \cdot \boldsymbol{r}_{uab} = 0$. The $\mathscr{L}_N = 0$ result directly follows from these conditions, and therefore the coherent state of N electrons is a stable equilibrium state. The proof of theorem 4 is thus complete; we showed the principal existence of coherent electrons states. Figure 1.6.1 illustrates this microscopic constellation of coherent electrons. We emphasize that the position of coherent electrons is stable only with respect to each other, but remains undetermined with respect to the laboratory frame: each delocalized electron forms a standing wave, and their positions are distributed within the standing wave.

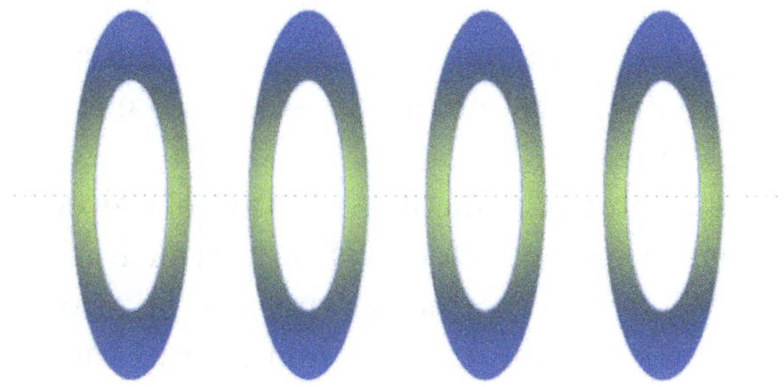

FIGURE 1.6.1. An illustration of electrons' coherent state. The dotted line represents the shared Zitterbewegung axis, the ellipses represent the 0.386 pm radius Zitterbewegung trajectories, and the colors represent the Zitterbewegung phase. Each electron has the same momentum, and their kinetic speed vectors point along the Zitterbewegung axis.

The above results directly show the absence of Larmor spin-precession in coherent electrons' equilibrium planes; the involved Zitterbewegung planes remain parallel to each other. Since all coherent electrons are part of the exact same quantum mechanical wave, individual electron spin measurement is impossible. The $\mathscr{L}_N = 0$ condition of equation 1.6.8 allows adding any number of electrons, and therefore we conclude that **coherent electrons obey Bose-Einstein statistics**.

Recalling that any electron's momentum is given by $e_a\boldsymbol{A}_a$, we can write down the Hamiltonian that corresponds to $\mathscr{L}_N$:

$$\mathscr{H}_N = \left[\sum_{a=1}^{N} e_a\boldsymbol{A}_a \cdot \boldsymbol{c}_a\right] - \mathscr{L}_N = \sum_{a=1}^{N} e_a\boldsymbol{A}_a \cdot \boldsymbol{c}_a - \left[e_a\boldsymbol{A}_a \cdot \boldsymbol{c}_a - e_aV_a\right] \tag{1.6.9}$$

$$\mathscr{H}_N = \sum_{a=1}^{N} e_aV_a$$

where we used equation 1.6.3 for evaluating $\mathscr{L}_N$.

The obtained Hamiltonian expression is rather simple. According to equation 1.6.7, V_a is minimized when the coherent electrons are as far apart as possible. The various constellations of inter-electron distances can be understood analogously to atomic orbitals: all atomic orbitals are equilibrium states, and the ground state orbital is the energy-minimizing equilibrium state. Therefore, in the energy-minimizing coherent state the N electrons maximize their distance - within the constraints of their enclosure.

A key result of the above analysis is that the inter-electron distances may vary while maintaining a coherent state. **This opens up a new degree of freedom for microscopic oscillations: the local density of coherent electrons can fluctuate**. Such oscillations will be investigated in the following chapter.

1.7. The incoherent-coherent electron state transition

1.7.1. The driving force of electrons' Bose-Einstein condensation. It follows from our results that the actual particle spin value plays no role, and that obeying Fermi-Dirac versus Bose-Einstein statistics is not any inherent property of a particle. Up to now, experimental investigations of Bose-Einstein condensates focused on low-temperature atomic condensates. In such atomic condensates, which form at ultra-low temperatures, experiments show that Bose-Einstein condensed atoms occupy the same quantum mechanical state, forming a matter-wave [**11, 12**] that is also spin-coherent [**13**]. Since the coherent state Lagrangian of section 1.6.4 universally applies to any particles that have a spin, the Bose-Einstein condensation of electrons is analogous to the atomic condensate case.

Let us consider a transition from the incoherent state of paired electrons into a variable state occupancy by coherent electrons. During such a transition, electrons enter one-by-one into a coherent state, when it is thermodynamically favorable. Within a delocalized standing wave, coherent electrons maximize their relative distance, as required by equation 1.6.9.

In vacuum, it follows from equation 1.6.9 that a coherent electron state is metastable because of the electrostatic repulsion term - unless these electrons are trapped in a potential well. Within a metal, the electrostatic electron potential is counter-balanced by the positively charged nuclear sites. It follows from equation 1.6.9 that a coherent state of conduction-band electrons has the same thermodynamic property as a non-interacting atomic gas.

Electron coherence therefore becomes driven by the energy gain of low-energy delocalized states being occupied by numerous electrons. However, the coherence of electrons can be disrupted by thermal fluctuations, and therefore coherent states remain stable only as long as they are thermodynamically favorable. In other words, Bose-Einstein statistics applies to delocalized electrons when the coherence equations of section 1.6.4 are thermodynamically favored over pair-wise spin correlations. The thermodynamic condition for electrons' Bose-Einstein condensation can be precisely calculated: delocalized electrons' Bose-Einstein condensation temperature shall be derived in chapter 3.

1.7.2. Overcoming the energy barrier against the break-up of incoherent spin-singlet pairs. Incoherent electrons' pairwise spin correlation, which forms spontaneously, involves local energy minimization via magnetic spin interactions. Larmor spin-precessing singlet electron pairs must be broken up before the involved electrons can synchronize their spins into a coherent state. While cooling a metal promotes electrons' Bose-Einstein condensation, at the same time it hinders the thermal break-up of Larmor spin-precessing singlet electron pairs. Whether a metal may become superconducting thus depends on the balance of these two factors.

In order to overcome the energy barrier against the break-up of incoherent spin-singlet pairs, some physical mechanism must provide the needed electron excitation energy. This cannot be thermal energy, since thermal energy fluctuations are negligible at low temperature. The remaining electron excitation source is the process that occasionally scatters conduction-band electrons; i.e. the same process that causes electric energy dissipation in a normal-state conductor provides the excitation energy to singlet electron pairs. As a consequence, a poorly conducting metal more easily enters superconducting state. The phenomenological Homes' law captures exactly this insight; it states that $T_c \sim n_s \sigma_{dc}^{-1}$, where T_c is the superconducting transition temperature, n_s is the density of superconducting electrons at $T \to 0$, and σ_{dc} is the DC electric conductivity just above T_c. Needless to say, the discovery of Homes' law surprised superconductivity researchers because nothing suggested in preceding theories that electron scattering would be beneficial for reaching the superconducting phase transition. For the first time, we shed light on the microscopic origin of Homes' law.

At ambient pressure, about half of elementary metals become superconducting, with T_c being in the 0-9 K temperature range. Under applied pressure, most elementary metals become superconducting, with T_c being in the 0-29 K temperature range. The exceptional elementary metals, which do not superconduct even under applied pressure are: i) ferromagnetic metals, ii) noble metals, and iii) several alkali metals. Noble and alkali metals are characterized by a relatively high electric conductivity; for the above explained reasons it is indeed anticipated that superconductivity would be most difficult to establish in these groups of metals.

Regarding ferromagnets, the interference of internal magnetic fields with Bose-Einstein condensation is well-known, and has been also observed in superfluids. This explains the absence of superconductivity in ferromagnets. Iron, for example, becomes superconducting above the pressure where its crystal structure changes into a non-ferromagnetic phase.

1.8. Summary

Our results demonstrate that it is the presence or absence of Larmor spin-precession in electron-electron interactions that determines the applicable statistics. A stable quantum mechanical wavefunction is an eigenstate solution of the Dirac equation, and such an eigenstate requires isotropic spin correlation of participating electrons. When Larmor precession is present, no more than two electrons may form isotropic spin correlation. In the absence of Larmor spin-precession, which is realized by coherent electrons, and arbitrary number of electrons may form isotropic spin correlation.

It follows from our results that the actual particle spin value plays no role, and that obeying Fermi-Dirac versus Bose-Einstein statistics is not any inherent property of a particle. Therefore the categorization of elementary particles into "fermion" versus "boson" classes is not always applicable.

The realization of Bose-Einstein condensed electron wavefunctions depends on the thermodynamics of a given system. Our analysis of coherent electron states demonstrates that they represent a lower entropy of electrons than the Larmor spin-precession involving incoherent states. These results facilitate a detailed energy change versus entropy change comparison of possible electron states, which is the pre-condition for a first-principles calculation of the Bose-Einstein transition temperature. This calculation will be the focus of a later chapter.

As the preference of Fermi-Dirac versus Bose-Einstein statistics is thermodynamically determined for a given system, an increasing mechanical pressure eventually becomes a thermodynamic driver for electron coherence: electron coherence mitigates the growing pressure by allowing the shared occupancy of low-energy states. The Bose-Einstein condensation of electrons should therefore play a role in geological and astrophysical phenomena that involve extreme gravitational forces and pressures.

Although we focused on electron examples, our equations in fact universally apply to any particle that has a spin. The Bose-Einstein condensation of electrons is observed in the few Kelvin to 100 Kelvin temperature range in superconductors. At ultra-low temperatures, the Bose-Einstein condensation of even ^{87}Rb nuclei is observed **[15]**, as anticipated. The temperature difference between electron versus ^{87}Rb condensation relates to the inverse proportionality between the particle mass and its Bose-Einstein condensation temperature.

If the Bose-Einstein condensation of some conduction band electrons is realized, it directly follows that a variable electron electron occupancy of the conduction band's ground state is superconducting due to the macroscopically large wavelength of added and removed electrons. The inter-electron distances may vary within such a multi-electron ground state, and this opens up a new degree of freedom for microscopic oscillations.

While the calculation of electron binding energies in the 1920's clarified one half of chemistry, it took 100 years to clarify the other half of chemistry which is related to electron statistics.

Appendix: Spin measurement along any direction

In section 1.5, we calculated spin correlations in the $x - z$ plane, according to the spin eigenvector definitions of table 2. Here, we show analogous calculations for the $y - z$ plane. Together, these results demonstrate the isotropy of spin correlations along any spin measurement direction. We start by expressing the $|+\rangle_z$ and $|-\rangle_z$ spin eigenvectors in terms of y-axis spin states:

$$|+\rangle_z = \begin{pmatrix} 1 \\ 0 \end{pmatrix} = \frac{1}{\sqrt{2}}\left[\begin{pmatrix} \frac{1}{\sqrt{2}} \\ \frac{i}{\sqrt{2}} \end{pmatrix} - i\begin{pmatrix} \frac{i}{\sqrt{2}} \\ \frac{1}{\sqrt{2}} \end{pmatrix}\right] = \frac{1}{\sqrt{2}}\left[|+\rangle_y - i\,|-\rangle_y\right]$$

$$|-\rangle_z = \begin{pmatrix} 0 \\ 1 \end{pmatrix} = \frac{1}{\sqrt{2}}\left[-i\begin{pmatrix} \frac{1}{\sqrt{2}} \\ \frac{i}{\sqrt{2}} \end{pmatrix} + \begin{pmatrix} \frac{i}{\sqrt{2}} \\ \frac{1}{\sqrt{2}} \end{pmatrix}\right] = \frac{1}{\sqrt{2}}\left[-i\,|+\rangle_y + |-\rangle_y\right]$$

The above equations mean that if we filter the electron to be in the $|+\rangle_z$ state, a subsequent spin measurement along the y axis yields equal probability of finding the electron in any of the two spin states. Similarly, upon filtering the electron to be in the $|-\rangle_z$ state, a subsequent spin measurement along the y axis yields again equal probability of finding the electron in any of the two spin states. This is analogous to the result explained in section 1.5 for the x direction; our mathematical notation captures the independence of spin measurements in orthogonal directions.

While in the $x - z$ plane a rotation of spin measurement direction was described by an ordinary rotation matrix, the above equations show that in the $x - z$ plane the spin eigenvectors are rotated according to the $R' = \begin{pmatrix} \cos\theta & -i\sin\theta \\ -i\sin\theta & \cos\theta \end{pmatrix}$ rotation matrix.

Moving on to electron spin correlations, consider the anti-parallel aligned spin-singlet state of two electrons. In terms of z-axis measurement, we described it in the following form:

$$|\uparrow\downarrow\rangle_z = \frac{1}{\sqrt{2}}\left[\begin{pmatrix} 1 \\ 0 \end{pmatrix} \otimes \begin{pmatrix} 0 \\ 1 \end{pmatrix} - \begin{pmatrix} 0 \\ 1 \end{pmatrix} \otimes \begin{pmatrix} 1 \\ 0 \end{pmatrix}\right]$$

For an arbitrary spin measurement direction in the $y - z$ plane, the eigenvectors transform via the above explained R' rotation matrix, and the reader may validate that the $\frac{1}{\sqrt{2}}\left[R'\begin{pmatrix} 1 \\ 0 \end{pmatrix} \otimes R'\begin{pmatrix} 0 \\ 1 \end{pmatrix} - R'\begin{pmatrix} 0 \\ 1 \end{pmatrix} \otimes R'\begin{pmatrix} 1 \\ 0 \end{pmatrix}\right]$ expression evaluates to the same result as the z-axis measurement. Equation 1.5.2 therefore expresses isotropic spin entanglement with respect to any measurement direction.

Lastly, consider the parallel aligned spin-triplet state of two electrons. In terms of z-axis measurement, we described it in the following form:

$$|\uparrow\downarrow\rangle_z = \frac{1}{\sqrt{2}}\left[\begin{pmatrix} 1 \\ 0 \end{pmatrix} \otimes \begin{pmatrix} 1 \\ 0 \end{pmatrix} + \begin{pmatrix} 0 \\ 1 \end{pmatrix} \otimes \begin{pmatrix} 0 \\ 1 \end{pmatrix}\right]$$

For an arbitrary spin measurement direction in the $y - z$ plane, the eigenvectors transform via the above explained R' rotation matrix, and the reader may validate that the $\frac{1}{\sqrt{2}}\left[R'\begin{pmatrix} 1 \\ 0 \end{pmatrix} \otimes R'\begin{pmatrix} 1 \\ 0 \end{pmatrix} + R'\begin{pmatrix} 0 \\ 1 \end{pmatrix} \otimes R'\begin{pmatrix} 0 \\ 1 \end{pmatrix}\right]$ expression evaluates to the same result as the z-axis measurement. Equation 1.5.3 therefore expresses isotropic spin entanglement with respect to any measurement direction.

Bibliography

[1] L. Ozyuzer et al "Emission of coherent THz radiation from superconductors", Science, volume 318.5854 (2007)

[2] J. Yang et al "Spin-triplet superconductivity in $K_2Cr_3As_3$", Science advances, volume 7.52 (2021)

[3] J. Nakamura et al "Direct observation of anyonic braiding statistics", Nature Physics, volume 16.9 (2020)

[4] P. O'Hara "Rotational Invariance and the Spin-Statistics Theorem", Foundations of Physics, volume 33. 9 (2003)

[5] P. O'Hara "A Generalized Spin-Statistics Theorem", Journal of Physics Conference Series, volume 845 (2017)

[6] A. Schweiger and G. Jeschke "Principles of Pulse Electron Paramagnetic Resonance" (2001)

[7] Y. Torizuka et al "Entangled pairs of 2p atoms produced in photodissociation of H_2 and D_2", Physical Review A , Volume 99.6 (2019)

[8] P. Milonni "The Quantum Vacuum", Academic Press (1994)

[9] J. L. Beck "Free Electron Paths from Dirac's Wave Equation Elucidating Zitterbewegung and Spin", arXiv:2506.20857 (2025)

[10] A. Kovacs et al "What is inside an electron: the physics of electromagnetic waves that are perceived as quantum mechanical waves", Zitter Institute (2025)

[11] W. Ketterle "Nobel lecture: When atoms behave as waves: Bose-Einstein condensation and the atom laser", Reviews of Modern Physics, Volume 72 (2002)

[12] T. Lahaye et al "The physics of dipolar bosonic quantum gases", Reports on Progress in Physics, Volume 72.12 (2009)

[13] K. Kim et al "Emission of spin-correlated matter-wave jets from spinor Bose-Einstein condensates", Physical Review Letters, Volume 127.4 (2021)

[14] D. Hestenes "Spin and uncertainty in the interpretation of quantum mechanics", American Journal of Physics, Volume 47 (1979)

[15] The Nobel Prize in Physics 2001. https://www.nobelprize.org/prizes/physics/2001/popular-information/

CHAPTER 2

The physical mechanism of perfect diamagnetism

Andras Kovacs[1] and **Giorgio Vassallo**[2,3]
[1] ExaFuse
[2] International Society for Condensed Matter Nuclear Science (ISCMNS)-UK.
[3] Università degli Studi di Palermo - Engineering Department, Palermo, Italy.
E-mail: andras.kovacs@broadbit.com

Abstract: We explore the physical origin of superconductors' perfect diamagnetism. It turns out that this diamagnetic effect differs from the dynamics of free-flowing electron pairs. The identified differences are highlighted; these are direct experimental refutations of "free-flowing electron pairs" based superconductivity models. Applying the condition of total energy minimization, we present the first rigorous derivation of the London equation, which governs this effect. We show how this diamagnetic effect enables the density measurement of superconducting electrons, calculate the microscopic electron vortices which collectively generate the Meissner flow of current, and show that the magnetic field of rotating superconductors is generated by the same diamagnetic effect.

2.1. Residual diamagnetism above T_c

Superconductors have perfect diamagnetism; they completely expel an externally applied magnetic field. Does superconductors' diamagnetism disappear abruptly or gradually above T_c? Figure 2.1.1 reveals a universal relationship of residual diamagnetism in superconducting materials, extending to twice the superconducting temperature in some cases. Therefore, diamagnetism does not just suddenly appear at T_c, but a residual diamagnetism remains even above T_c.

It can be seen in figure 2.1.1 that the residual diamagnetism of elementary metals, such as Pb, extends only very slightly above the critical temperature. In contrast, T_{onset} extends far above T_c in high-temperature superconductors.

2.2. The London equation

The London equation was proposed in the 1930s as a phenomenological equation for describing the interaction between magnetic fields and superconductors, which are known for their perfect diamagnetism. One may try to understand such interaction by using Maxwell's equation to calculate the currents induced by a time-varying magnetic field, which is gradually ramped up from zero to its final value. Such calculation is presented for example in chapter 6 of **[2]**. The calculation of induced currents reveals that a perfect conductor indeed diamagnetically responds to a time-varying magnetic field; the induced surface currents expel the magnetic field from its interior. Unfortunately, such calculation does not yet explain how diamagnetism works in superconductors. As illustrated in figure 2.2.2, a superconductor expels magnetic fields even when $\frac{dB}{dt} = 0$; e.g. when its temperature is cooled below T_c. A more general approach is to solve for the global minimum of free energy, and such approach is applicable for Bose-Einstein condensates. A recent textbook recognizes that the London equation gives an energy minimizing solution **[3]**, but omits a proper derivation. This same recognition appears in

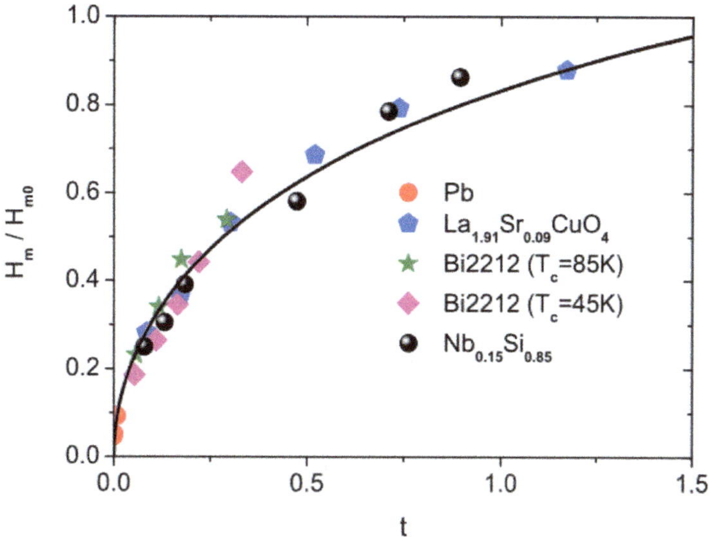

FIGURE 2.1.1. Above-T_c residual diamagnetism in various superconductors. The horizontal axis measures $t = \frac{T}{T_c} - 1$, and thus $T = T_c$ at the origin. The vertical axis shows magnetization H_m at a given temperature, as a fraction of the room temperature magnetization H_{m0}. The zero of the vertical axis represents perfect diamagnetism. Reproduced from [**1**].

several articles [**4, 5**], but always without mathematical derivation, as far as the authors know. We therefore endeavor to derive the London equation in the following paragraphs.

First counter-evidence to the "free-flowing electron pairs" based superconductivity models: We firstly consider type-I superconductors, which have isotropic structure. We note that a series of works [**6, 7, 8, 9**] demonstrated already in the 1930s that the superconducting-normal state transition in the presence of a magnetic field occurs without sudden energy dissipation; calorimetry indicates that the superconducting current stops without any heat spike. Figure 2.2.1 illustrates the $B - T$ phase diagram of a type-I superconductor. Starting from state "P" on figure 2.2.1, a reversible superconducting-normal state transition may be induced either by raising temperature or by raising magnetic field strength. According to the traditional theory of superconductivity, superconducting electrons flow without energy dissipation, and they are abruptly converted to normal conduction band electrons once the external magnetic field reaches a critical value. Thus, traditional theory predicts that the superconducting "Meissner flow" inevitably produces Joule heating upon superconducting-to-normal phase transition due to the transition of superconducting electrons into normal state, that flow with dissipation. The predicted Joule heating, which should be generated during the phase transition, will make it an irreversible transition. But this prediction contradicts the experimentally observed reversible phase transition in a magnetic field. This 90 years old contradiction indicates that serious revisions are needed in the standard theory.

The superconducting "Meissner flow", which causes the perfect diamagnetism of superconductors, comprises induced currents of Bose-Einstein condensed electrons. The microscopic structure of this flow shall be described in the last part of this chapter. The Meissner flow is illustrated in figure 2.2.2, and it carries an associated kinetic energy. What happens to superconducting electrons' kinetic energy beyond the superconducting-to-normal phase transition in a type-I superconductor? The example of Pb on figure 2.1.1 already revealed that there is no abrupt change of diamagnetism, and therefore the kinetic energy remains nearly the same also on the normal side of the phase transition.

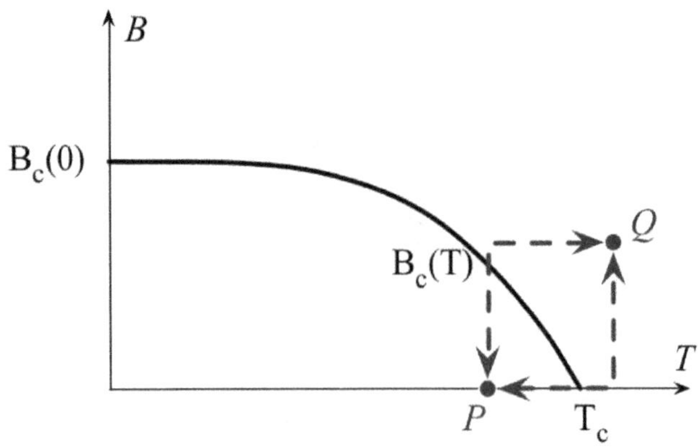

FIGURE 2.2.1. The phase diagram for a type I superconductor. B is the externally applied magnetic field, and T is the temperature. In the figure, state "P" is in the superconducting phase and state "Q" is in the normal phase. There is a reversible state change between "P" and "Q", irrespective of the path taken between them.

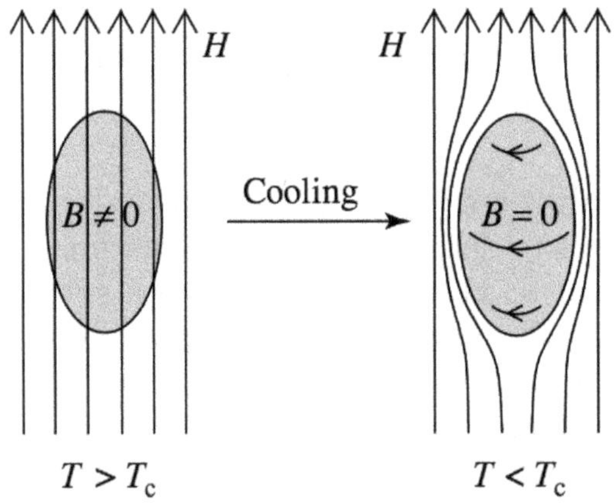

FIGURE 2.2.2. An illustration of the Meissner flow, which arises below T_c and expels magnetic field from the superconductor's interior. Horizontal arrows indicate Meissner flow currents in the superconductor's surface region.

According to figure 2.1.1, the only change is that the perfect diamagnetism of a Bose-Einstein condensate is replaced by a residual diamagnetism. In the reverse direction, this residual diamagnetism converges to perfect diamagnetism as temperature is lowered to T_c. Perfect diamagnetism is the vertical axis' zero value on figure 2.1.1. It is seen from figure 2.1.1 that there is a smooth convergence to perfect diamagnetism, which means that the system is in a thermodynamic equilibrium at any point, and thus state changes are reversible. One may also say that the application of external magnetic field lowers the effective T_c. As the magnetic field strength is raised from state P to B_c, the temperature at B_c indeed becomes the new T_c: any further increase of temperature triggers superconducting-to-normal phase transition.

Let us consider what thermodynamic equilibrium looks like in a superconducting material, when it is in superconducting phase. Perfect diamagnetism means that magnetic energy density is zero in the interior. In contrast, there is a magnetic energy density ϱ_m in the exterior, while superconducting electrons' translational kinetic energy density ϱ_k

is zero in that region. The challenge is to understand what happens in the transitionary surface region between the exterior and interior zones, where ϱ_m and ϱ_k are both non-zero.

Anywhere in the surface region, magnetic energy density is given by:

$$\varrho_m = \frac{1}{2\mu_0}\boldsymbol{B}^2 \tag{2.2.1}$$

The kinetic energy density of superconducting electrons is:

$$\varrho_k = \frac{m_{eff}}{2}\boldsymbol{v}_s^2 n_s \tag{2.2.2}$$

where n_s is the density of superconducting electrons and $\boldsymbol{v}_s$ is their velocity. The question arises whether it is correct to use m_{eff} in the above equation, rather than the bare electron mass. Superconducting electrons are in the lowest energy of the conduction band. Electrons occupying the lowest-energy conduction band states are observed in semiconductors, and experiments confirm that m_{eff} applies to delocalized electrons in semiconductors. The use of m_{eff} is therefore justified. The corresponding current can be written as:

$$\boldsymbol{J} = e\boldsymbol{v}_s n_s$$

Maxwell's equation defines the relationship between $\boldsymbol{J}$ and $\boldsymbol{B}$:

$$\mu_0\boldsymbol{J} = \nabla \times \boldsymbol{B}$$

Substituting the above two equations into 2.2.2, we express kinetic energy density in terms of magnetic field:

$$\varrho_k = \frac{m_{eff}}{2e^2\mu_0^2 n_s}\left(\nabla \times \boldsymbol{B}\right)^2 \tag{2.2.3}$$

Using equations 2.2.1 and 2.2.3, we can express the total energy density as a function of magnetic field:

$$\varrho\left(\boldsymbol{B}\right) = \varrho_m + \varrho_k = \frac{1}{2\mu_0}\boldsymbol{B}^2 + \frac{m_{eff}}{2e^2\mu_0^2 n_s}\left(\nabla \times \boldsymbol{B}\right)^2 \tag{2.2.4}$$

Suppose there is a disturbance or variation of the magnetic field; we denote this magnetic field variation as vector field $\boldsymbol{G}$. The derivative of energy density with respect to infinitesimal magnetic field variations is:

$$\varrho'\left(\boldsymbol{B},\boldsymbol{G}\right) = \frac{\partial}{\partial\varepsilon}\varrho\left(\boldsymbol{B} + \varepsilon\boldsymbol{G}\right)|_{\varepsilon\to 0} \tag{2.2.5}$$

We now evaluate the above differential by substituting equation 2.2.4:

$$\varrho'\left(\boldsymbol{B},\boldsymbol{G}\right) = \frac{1}{\mu_0}\left(\boldsymbol{B}\cdot\boldsymbol{G}\right) + \frac{m_{eff}}{e^2\mu_0^2 n_s}\left(\nabla \times \boldsymbol{B}\right)\cdot\left(\nabla \times \boldsymbol{G}\right) \tag{2.2.6}$$

where we obtained the last term by switching the order of the two differential operators:

$$\frac{\partial}{\partial\varepsilon}\left(\nabla \times \left(\boldsymbol{B} + \varepsilon\boldsymbol{G}\right)\right) = \nabla \times \frac{\partial}{\partial\varepsilon}\left(\boldsymbol{B} + \varepsilon\boldsymbol{G}\right) = \nabla \times \boldsymbol{G}$$

Thermodynamic equilibrium requires that the total system energy remains invariant with respect to small changes in the magnetic field, i.e. the system occupies the lowest energy state:

$$\int_{\Omega} \varrho' (\boldsymbol{B}, \boldsymbol{G}) \; dV = 0$$

for any choice of $\boldsymbol{G}$, keeping in mind that the variational field $\boldsymbol{G}$ vanishes at the boundary. Here, Ω is the volume region that contains our system. Substituting equation 2.2.6 into the above equilibrium requirement, we obtain:

$$\int_{\Omega} \left[\frac{1}{\mu_0} (\boldsymbol{B} \cdot \boldsymbol{G}) + \frac{m_{eff}}{e^2 \mu_0^2 n_s} (\nabla \times \boldsymbol{B}) \cdot (\nabla \times \boldsymbol{G}) \right] \; dV = 0$$

We use the $\int_\Omega (\nabla \times \boldsymbol{B}) \cdot (\nabla \times \boldsymbol{G}) \; dV = \int_\Omega (\nabla \times (\nabla \times \boldsymbol{B})) \cdot \boldsymbol{G} \, dV$ identity, which is proven in the appendix. Note, that we substituted $\nabla \times \boldsymbol{B}$ for the vector field $\boldsymbol{F}$ that is used in the appendix. The above equation can therefore be written as:

$$\int_{\Omega} \left[\frac{1}{\mu_0} \boldsymbol{B} + \frac{m_{eff}}{e^2 \mu_0^2 n_s} \nabla \times (\nabla \times \boldsymbol{B}) \right] \cdot \boldsymbol{G} \, dV = 0$$

Since the above equation must remain zero for any choice of $\boldsymbol{G}$, the equilibrium condition is given by:

$$\boldsymbol{B} + \frac{m_{eff}}{\mu_0 n_s e^2} \nabla \times (\nabla \times \boldsymbol{B}) = 0 \tag{2.2.7}$$

which is known as the London equation. Our simple derivation shows that the London equation solves for the lowest energy state.

Since magnetic fields have zero divergence, we get $\nabla \times (\nabla \times \boldsymbol{B}) = -\nabla^2 \boldsymbol{B}$, and the London equation may be written as:

$$\nabla^2 \boldsymbol{B} = \frac{\mu_0 n_s e^2}{m_{eff}} \boldsymbol{B} \tag{2.2.8}$$

The solution of the above equation is a $\boldsymbol{B} = \boldsymbol{B}_0 e^{-x/\lambda_L}$ type exponential decay of $\boldsymbol{B}$ from the surface towards the interior, over the characteristic distance λ_L :

$$\frac{1}{\lambda_L} = \sqrt{n_s \frac{\mu_0 e^2}{m_{eff}}} \tag{2.2.9}$$

Thus we obtain a relationship between λ_L and n_s:

$$\frac{1}{\lambda_L} \sim \sqrt{n_s} \tag{2.2.10}$$

Summarizing our results up to now, we demonstrated that superconducting Meissner flows are incompatible with free-flowing electron pair models of superconductivity, including incompatibility with the phonon-mediated superconductivity model. To be best of our knowledge, we obtained for the first time a mathematically complete derivation of the London equation.

We apply equation 2.2.9 to calculate the n_s value of elementary metals. Lead metal is a historically most studied superconductor. It has T_c=7.2 K and $m_{eff} = 1.9 m_e$. Its 0.34 nm inter-nuclear distance implies n=(0.27 nm)$^{-3}$ delocalized electron density with +2 valence state of nuclear sites. Applying these values to equation 2.2.9 we obtain λ_L (0)=33 nm, which is the same order of magnitude as the experimentally measured

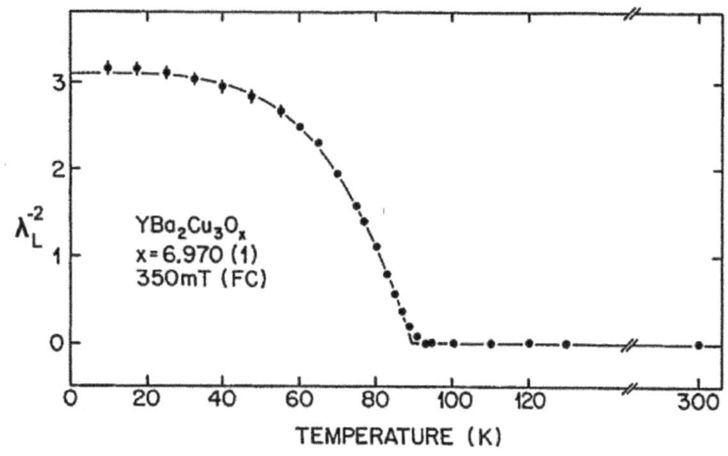

FIGURE 2.2.3. The chart of λ_L^{-2} in $YBa_2Cu_3O_{6+x}$ (T_c=92 K), as a function of temperature. λ_L^{-2} is evaluated via muon spin relaxation measurement, reproduced from [**10**].

$\lambda_L(0)$=52 nm value. This result means that at $T \to 0$ most conduction band electrons condense into the superconducting ground state.

We apply equation 2.2.9 to calculate the n_s value of high-temperature superconductors. K_3C_{60} is an isotropic superconductor, and its conduction band electron density is defined by its K^+ spacing. K_3C_{60} has $\lambda_L(0)$=480 nm at as $T \to 0$ [**11**] and $m_{eff} = 2.4m_e$ [**12**] values. Applying these values to equation 2.2.9 we obtain $n_s(0)$=(1.5 nm)$^{-3}$, which again corresponds quite well to the 1.4 nm distance between neighboring K^+ sites.

So far we studied the diamagnetism of isotropic materials. Most high-temperature superconductors are however 2-dimensional layered materials. Although magnetic fields are expelled also from such layered superconductors, regardless of their orientation, λ_L is no longer isotropic: its value is several times larger orthogonally to the 2-dimensional metallic planes than in parallel directions. Nevertheless, our derivation of equations 2.2.7 and 2.2.9 continues to be valid for layered materials: the employed variational method does not require isotropy.

As illustrated in figure 2.2.3, $\lambda_L \to \infty$ as $T \to T_c$, and this implies that the density of superconducting electrons is depleted near T_c according to equation 2.2.9. Keeping in mind that $T_c = T_{BEC}$, we get $n_s \to 0$ as $T \to T_{BEC}$.

2.3. Rotating superconductors

When a superconducting cylinder is being rotated, a magnetic field appears in its interior. Traditionally, this magnetic field was thought to originate from the inertia of free-flowing electrons. The phenomenology of rotating superconductors is thoroughly reviewed in reference [**13**].

Measurements show that the uniform magnetic field which fills all the available interior has $B = -\frac{2m_e}{e}\omega$ strength, where ω is the angular frequency of the cylinder's rotation. Normally, a magnetic flux is generated by a current $I = dQ/dt$, and at first it seems obvious that the magnetic field would be generated by a current caused by the inertia of free-flowing electrons, that do not follow the rotation of the superconductive body. It is interesting to note that the above formula does not depend on the number of charge carriers involved in the generation of the magnetic field B, but depends on the electron mass m_e. In current textbooks, this is the end of the story about rotating superconductors.

Let us consider what happens to hypothetical free-flowing electrons at the start of rotation. Since the already superconducting electrons are supposedly completely free to

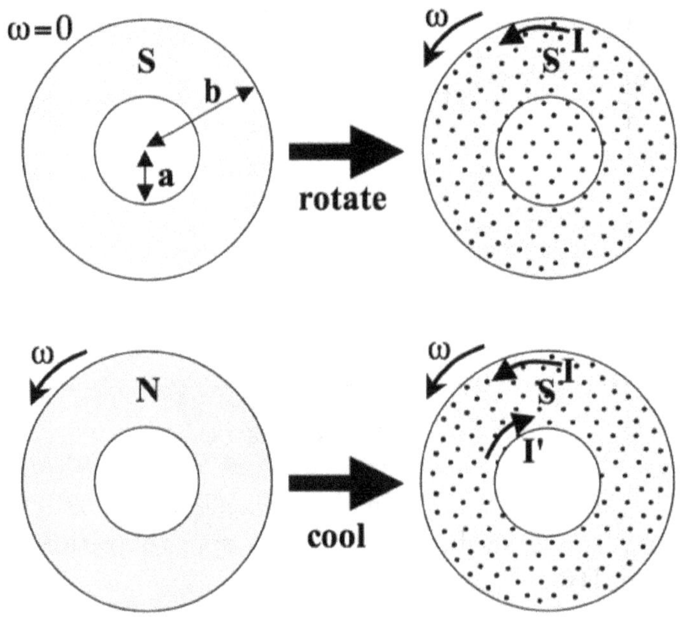

FIGURE 2.3.1. The rotating and cooling of a superconducting hollow cylinder. Top: cylinder cooled into the superconducting state and then set into rotation. Bottom: cylinder set into rotation while in the normal state and then cooled into the superconducting state. Dots represent magnetic field lines which are perpendicular to the figure, "N" represents normal state, and "S" represents superconducting state. Reproduced from [**13**].

flow in the body, their inertia keeps them at the same location at the moment when the body starts to rotate. During this start of rotation, a time-dependent magnetic field is generated by the diverging motion of positive ions and superconducting electrons, that induces a Faraday electric field pushing the superconducting electrons in the same direction as the ions. Consequent to this induced Faraday electric field, the superconducting electrons attain the same velocity as the ions in the interior of the cylinder but lag slightly behind in the outer surface layer, thus giving rise to the static magnetic field configuration shown at the top of figure 2.3.1.

Second counter-evidence to the "free-flowing electron pairs" based superconductivity models: As discussed in section 2.2, the dynamics of superconducting electron currents are dictated by the principle of total energy minimization. Therefore, in the context of a free-flowing electrons based model, it should be possible to derive the $\boldsymbol{B} = -\frac{2m_e}{e}\boldsymbol{\omega}$ formula from the London equation. The London equation's scaling factor, which is given by equation 2.2.9, implies $\lambda_L \sim \sqrt{\frac{m_e}{n_s}\frac{1}{e}}$ relationship. There is no obvious way to eliminate the dependence on superconducting electron density, as needed to arrive at the $B = -\frac{2m_e}{e}\omega$ formula. Nevertheless, reference [**14**] claims to derive this formula from the London equation, and it lists several studies which contain similar derivations. All these works struggle with the challenge of reconciling the $B = -\frac{2m_e}{e}\omega$ formula with the London equation. The problem is very deep-rooted. Consider a long superconductive cylinder, which is rotating. The magnetic field in its interior is $\boldsymbol{B} = -\frac{2m_e}{e}\boldsymbol{\omega}$. We cut the cylinder into two shorter cylinders. Each rotating cylinder still has $\boldsymbol{B} = -\frac{2m_e}{e}\boldsymbol{\omega}$ magnetic field in its interior, and it retains the same value regardless of whether the cylinders are on top of each other or far apart. But this situation is totally contrary to the basic principles of current-generated magnetic fields, which dictate that magnetic fields must add up when the two pieces are placed on top of each other - just like magnetic fields of solenoid segments add up. We thus come to the following conclusion: the magnetic field of a rotating superconductor cannot be generated by free-flowing currents! This conclusion is further corroborated by the situation depicted in the bottom half of figure 2.3.1. In that scenario, the free-flowing current model dictates that all superconducting electrons lag behind the main body rotation, but the experiment shows inner surface superconducting electrons rotating faster than the main body. The traditional explanations of the free-flowing current model, which are electron inertia based, break down once again.

The correct analysis of rotating superconductors must take into account the consequences of a non-inertial rotating reference frame; this analysis can be found in reference **[15]**. Its author finds that in a rotating reference frame a charged particle responds to the electromagnetic field according to the following equation of motion:

$$m_e \boldsymbol{a} = e\left(\boldsymbol{E}' + \boldsymbol{v} \times \boldsymbol{B}'\right)$$

where:

$$\boldsymbol{E}' = \boldsymbol{E} + \frac{m_e}{e}\boldsymbol{\omega} \times (\boldsymbol{\omega} \times \boldsymbol{r})$$

$$\boldsymbol{B}' = \boldsymbol{B} + \frac{2m_e}{e}\boldsymbol{\omega}$$

In the above expressions $\boldsymbol{\omega}$ denotes the angular speed of rotation at radial distance $\boldsymbol{r}$, while $\boldsymbol{E}$ and $\boldsymbol{B}$ denote the electric and magnetic fields in the laboratory frame.

The extra terms of the electric and magnetic field formulas correspond to the usual centrifugal and Coriolis forces, associated with motion in a rotating frame. The superconducting Meissner effect then ensures that the effective $\boldsymbol{B}'$ field is zero. This $\boldsymbol{B}' = 0$ condition of the rotating frame directly leads to the following result in the laboratory frame:

$$\boldsymbol{B} = -\frac{2m_e}{e}\boldsymbol{\omega} \tag{2.3.1}$$

Equation 2.3.1 is known as the London moment formula. To verify the correctness of the above formulas, let us consider an experimental evidence for the $\boldsymbol{B}' = \frac{2m}{e}\boldsymbol{\omega}$ pseudo-field. The authors of **[16]** measured the ^{13}C precession frequency in materials rotated at various angular frequencies. The nuclear precession frequency is directly proportional to the perceived magnetic field; this experiment thus measures the magnetic field that a particle perceives in a rotating reference frame. As can be seen in figure 2.3.2, the outcome of this experiment confirms the calculations of reference **[15]**.

Our derivation of the $\boldsymbol{B} = -\frac{2m_e}{e}\boldsymbol{\omega}$ magnetic field in the interior of rotating superconductors resolves the long-standing puzzle of why this formula does not depend on the superconducting electron density and why it does not depend on the length of the rotating cylinder either. Having clarified that the magnetic field of rotating superconductors arises from the same Meissner effect that is responsible for superconductors' diamagnetism in the laboratory frame, what remains is to understand the Meissner current's microscopic structure.

2.4. The microscopic structure of Meissner flows

We showed that the London equations describe an energy-minimizing equilibrium solution within the superconductor's surface region. The Meissner flow of superconducting electrons, which causes superconductors' perfect diamagnetism, carries kinetic energy. The aim of this section is to understand the microscopic structure of Meissner flows, i.e. to identify the corresponding quantum mechanical state.

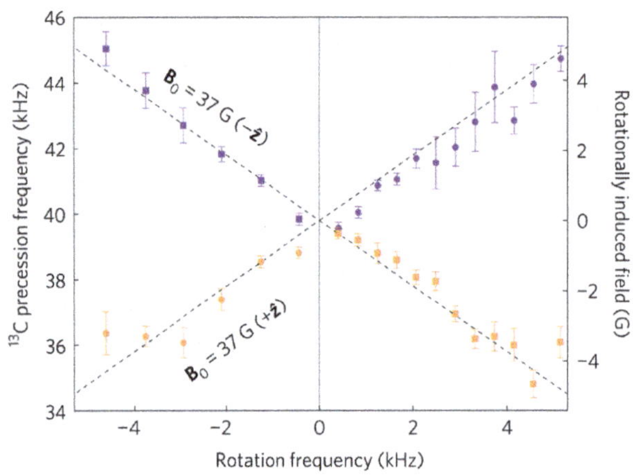

FIGURE 2.3.2. A direct evidence of the $\boldsymbol{B}' = \frac{2m}{e}\boldsymbol{\omega}$ pseudo-field, measured via the ^{13}C precession frequency. Reproduced from [**16**].

Third counter-evidence to the "free-flowing electron pairs" based superconductivity models: Up to now, Meissner flows were modeled as zero-resistance perimetric flows along the superconductor's surface. However, a perimetric Meissner flow model leads to physical paradoxes. To see this, let us take the example of a round shaped superconductor illustrated in figure 2.2.2. The presumed circular flow of superconducting electrons requires their acceleration towards the center. Such electron acceleration implies a continuous radiation of electromagnetic energy, even if superconducting electrons were free-flowing, and this would eventually stop the Meissner flows. Experiments however indicate that superconducting Meissner flows do not stop even after several years.

Since superconducting Meissner flows do not radiate, superconductors' diamagnetism must be generated by microscopic circular flows; these flows correspond to some quantum mechanical ground state which can not loose kinetic energy. Figure 2.4.1 illustrates this microscopic Meissner flow structure. In the following, we calculate the radius of such Meissner flow vortexes.

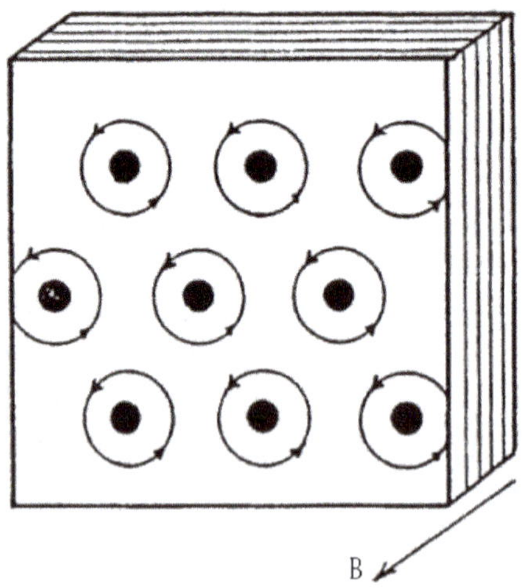

FIGURE 2.4.1. An illustration of vortexes that comprise the Meissner flow. The externally applied magnetic field is represented by the arrow.

One may ask why figure 2.4.1 does not apply to ordinary metals. In ordinary metals, the incoherent Fermi sea wavefunctions comprise standing wave modes; these standing waves do not have any degree of freedom for a microscopic oscillation shown in figure 2.4.1.

As discussed at the end of section 1.6, coherent electron states are occupied by a large number of electrons, and their inter-electron distances may vary while maintaining their coherent state. This opens up a new degree of freedom for microscopic oscillations: the density of coherent electrons can fluctuate around positively charged nuclei. A microscopic circular flow can be modeled as an oscillation in the x and y directions, with $\frac{\pi}{2}$ phase between them. Electrons' oscillation around fixed positive charges is a so-called Langmuir oscillation, and its angular frequency ω_L is derived as follows. Suppose that the density of superconducting electrons has a variation δn_s from the mean, and the average flow speed of these electrons is v. The continuity condition along the x direction becomes:

$$\frac{\partial(\delta n_s)}{\partial t} = -n_s \frac{\partial v}{\partial x} \tag{2.4.1}$$

Taking the time differential of the above equation, we obtain:

$$\frac{\partial^2(\delta n_s)}{\partial t^2} = -n_s \frac{\partial}{\partial t}\frac{\partial v}{\partial x} = -n_s \frac{\partial}{\partial x}\frac{\partial v}{\partial t} \tag{2.4.2}$$

The electron density variation δn_s generates an electric field E, according to Poisson's equation:

$$\frac{\partial E}{\partial x} = -\frac{e}{\varepsilon_0}\delta n_s \tag{2.4.3}$$

The generated E field changes the momentum of superconducting electrons:

$$m_{eff}\frac{\partial v}{\partial t} = -eE \tag{2.4.4}$$

where m_{eff} is the effective electron mass. Substituting equations 2.4.3-2.4.4 into 2.4.2, we obtain a wave equation for δn_s:

$$\frac{\partial^2(\delta n_s)}{\partial t^2} = \frac{e n_s}{m_{eff}}\frac{\partial E}{\partial x} = -\frac{e^2 n_s}{\varepsilon_0 m_{eff}}\delta n_s \tag{2.4.5}$$

This wave equation is solved by the oscillatory motion of superconducting electrons at the following angular frequency:

$$\omega_L = \sqrt{\frac{e^2 n_s}{\varepsilon_0 m_{eff}}}$$

An analogous analysis for the y direction yields the same frequency. This angular frequency ω_L is the natural oscillation frequency of superconducting electrons. We notice that ω_L can be expressed in terms of the London penetration depth parameter, that is given by equation 2.2.9:

$$\omega_L = \sqrt{c^2}\sqrt{n_s \frac{\mu_0 e^2}{m_{eff}}} = \frac{c}{\lambda_L} \tag{2.4.6}$$

This remarkable relationship demonstrates that superconducting electrons' kinetic energy density, which was denoted as ϱ_k in section 2.2, corresponds to their Langmuir oscillation.

These oscillations in the x and y directions result in a vortex motion of superconducting electrons, which generates their diamagnetic response. Since this vortex is in a quantum mechanical ground state, each involved electrons' angular momentum is quantized to $\hbar$:

$$m_{eff} r_{\circlearrowleft} \left(r_{\circlearrowleft} \omega_L \right) = \hbar \tag{2.4.7}$$

where $r_{\circlearrowleft}$ is the Meissner vortex radius illustrated in figure 2.4.1. Such angular momentum quantization is essentially a continuity condition on superconducting electrons' wavefunction along the vortex circle. I.e. the $\hbar$ angular momentum condition corresponds to a non-radiating ground state. We can now solve for $r_{\circlearrowleft}$:

$$r_{\circlearrowleft} = \sqrt{\frac{\hbar}{m_{eff}\omega_L}} = \sqrt{\frac{\hbar \lambda_L}{m_{eff} c}} \tag{2.4.8}$$

Using the above result, one may evaluate $r_{\circlearrowleft}$ for various superconducting materials. Since λ_L grows with increasing temperature, r_M also grows with increasing temperature.

Lead material has $\lambda_L\,(0)$=52 nm, and taking $m_{eff} = 1.9 m_e$ yields $r_{\circlearrowleft}$=0.1 nm at low temperatures. The $YBa_2Cu_3O_{6.95}$ material, which is a representative high-temperature superconductor, has $\lambda_L\,(0)$=140 nm and $m_{eff} \approx 2.2 m_e$, which yields $r_{\circlearrowleft}$=0.16 nm at low temperatures. At low temperatures, the $2r_{\circlearrowleft}$ diameter has a similar size to the unit cell dimension of these materials. These examples demonstrate that at $T \to 0$ a Meissner vortex diameter is similar to the unit cell size. This makes sense because these microscopic Langmuir oscillations are centered around the positively charged Pb nuclei in lead, and around the positively charged Cu nuclei in $YBa_2Cu_3O_{6.95}$.

Taking an example with larger $\lambda_L\,(0)$ value, K_3C_{60} has $\lambda_L\,(0)$=480 nm and $m_{eff} = 2.4 m_e$ values. Applying these values to equation 2.4.8 we obtain $r_{\circlearrowleft}$=0.28 nm at low temperatures.

The above-calculated low-temperature $r_{\circlearrowleft}$values grow larger with increasing temperatures, as defined by the equation 2.4.8. When the temperature is raised towards T_c, the $r_{\circlearrowleft}$ values grow into the nanometer range.

2.5. Conclusions

We explored the physical origin of superconductors' perfect diamagnetism. The Meissner effect is generated by circular oscillations of Bose-Einstein condensed electrons around positively charged lattice sites, with ~0.1 nm mean radius at low temperatures. Such oscillations are a quantum mechanical version of Langmuir oscillations, and they arise because their presence minimizes the total energy density. We showed that these oscillations are non-radiating because they correspond to a quantum mechanical ground state. We demonstrated that rotating superconductors generate magnetic field through the same physical process, as it minimizes the total energy density in the rotating reference frame. A review of related experimental data demonstrates that the Meissner effect is incompatible with "free-flowing electron pairs" based models, which invalidates preceding superconductivity theories. For the first time, we worked out a precise derivation of the London equation. We also showed how the diamagnetic Meissner effect enables the density measurement of superconducting electrons.

Acknowledgements: The authors thank Jan von Pfaler for his help with the London equation's mathematical derivation.

Appendix: a useful vector field identity

Let $\boldsymbol{F}, \boldsymbol{G}$ be vector fields, and let Ω be a volume region in space. Let σ denote the surface of Ω. The field $\boldsymbol{G}$ will serve as "variation field", and therefore we require it to vanish on the surface of Ω: i.e. $\boldsymbol{G}\,|_\sigma = 0$. We start from the following well-known vector field identity:

$$\nabla \cdot (\boldsymbol{F} \times \boldsymbol{G}) = (\nabla \times \boldsymbol{F}) \cdot \boldsymbol{G} - \boldsymbol{F}\,(\nabla \times \boldsymbol{G})$$

We integrate the above equation over Ω:

$$\int_\Omega \nabla \cdot (\boldsymbol{F} \times \boldsymbol{G})\; dV = \int_\Omega (\nabla \times \boldsymbol{F}) \cdot \boldsymbol{G}\, dV - \int_\Omega \boldsymbol{F}\,(\nabla \times \boldsymbol{G})\; dV$$

Upon rearranging terms, we get:

$$\int_\Omega \boldsymbol{F}\,(\nabla \times \boldsymbol{G})\; dV = \int_\Omega (\nabla \times \boldsymbol{F}) \cdot \boldsymbol{G}\, dV - \int_\Omega \nabla \cdot (\boldsymbol{F} \times \boldsymbol{G})\; dV \tag{2.5.1}$$

We evaluate the last term of the above equation by changing from volume integration to bounding surface integration:

$$\int_\Omega \nabla \cdot (\boldsymbol{F} \times \boldsymbol{G})\; dV = \int_\sigma \boldsymbol{F} \times \mathbf{G}\, dA = 0$$

The above equation evaluates to zero because $\boldsymbol{G}$ vanishes on the surface of Ω. Therefore, equation 2.5.1 simplifies to:

$$\int_\Omega \boldsymbol{F}\,(\nabla \times \boldsymbol{G})\; dV = \int_\Omega (\nabla \times \boldsymbol{F}) \cdot \boldsymbol{G}\, dV \tag{2.5.2}$$

Bibliography

[1] T. Schneider et al "Diamagnetism, Nernst signal, and finite-size effects in superconductors above the transition temperature", Physical Review B, Volume 83.14 (2011)

[2] J. E. Hirsch "Superconductivity begins with H", World Scientific (2020)

[3] K. Fossheim et al "Superconductivity - Physics and Applications", John Wiley & Sons (2004), Chapter 6.1.2

[4] A. Badia-Majos "Understanding stable levitation of superconductors from intermediate electromagnetics", American journal of physics, Volume 74.12 (2006)

[5] H. Koizumi et al "Theory of supercurrent in superconductors", International Journal of Modern Physics B, Volume 34.31 (2020)

[6] W. Keesom et al "Measurements of the latent heat of thallium connected with the transition, in a constant external magnetic field, from the supraconductive to the non-supraconductive state", Physica, Volume 1 (1934)

[7] W. Keesom et al "Measurements of the latent heat of tin in passing from the supraconductive to the non-supraconductive state", Physica, Volume 3 (1936)

[8] W. Keesom et al "Measurements of the latent heat of tin while passing from the superconductive to the non-superconductive state at constant temperature", Physica, Volume 4 (1937)

[9] W. Keesom et al "On the reversibility of the transition process between the superconductive and the normal state", Physica, Volume 5 (1938)

[10] H. Keller "Positive muons as probes in high-T_c superconductors", in "Exotic Atoms in Condensed Matter", Springer (1992)

[11] Y. J. Uemura et al "Magnetic-field penetration depth in K_3C_{60} measured by muon spin relaxation", Nature, Volume 352.6336 (1991)

[12] Y. T. Iwasa et al "Optical properties of C_{60} and K_3C_{60}", Synthetic metals, Volume 56.2-3 (1993)

[13] J. E. Hirsch "Defying inertia: how rotating superconductors generate magnetic fields", arXiv:1812.06780v2 (2019)

[14] H. Essén et al "The Darwin-Breit magnetic interaction and superconductivity", arXiv:1312.1607 (2013)

[15] B. R. Holstein "The Aharonov-Bohm effect and variations", Contemporary Physics, Volume 36.2 (1995)

[16] A. A. Wood et al "Magnetic pseudo-fields in a rotating electron–nuclear spin system", Nature Physics, Volume 13.11 (2017)

CHAPTER 3

The thermodynamics of electrons' Bose-Einstein condensation

Andras Kovacs[1]
[1] ExaFuse
E-mail: andras.kovacs@broadbit.com

ABSTRACT. Bose-Einstein condensation is an intensely studied quantum phenomenon that emerges at low temperatures. While preceding Bose-Einstein condensation models do not consider what statistics applies above the condensation temperature, we show that neglecting this question leads to inconsistencies. A mathematically rigorous calculation of Bose-Einstein condensation temperature requires evaluating the thermodynamic balance between coherent and incoherent particle populations. The first part of this work develops such an improved Bose-Einstein condensation temperature calculation, for both three-dimensional and two-dimensional scenarios. The progress over preceding Bose-Einstein condensation models is particularly apparent in the two-dimensional case, where preceding models run into mathematical divergence.

In the second part of this chapter, we validate our mathematical model against experimental superconductivity data. A remarkable match is found between experimental data and the calculated Bose-Einstein condensation temperature formulas. Being able to calculate electrons' Bose-Einstein condensation dynamics therefore facilitates a rational search for higher-temperature superconductors.

3.1. Introduction

The theory of Bose-Einstein statistics was formulated nearly 100 years ago, and the corresponding Bose-Einstein condensation temperature formula was calculated a few decades later for the isotropic gas scenario [**3**]. Experimentally, the first undisputed observation of Bose-Einstein condensation was reported for ultracold dilute gases in 1995 [**1**].

On the other hand, superconductivity has been also intensely studied during the past 100 years. Since electric resistivity originates from scattering events, such as electron scattering on crystal defect sites or electron-electron scattering, superconducting electrons must have the ability to pass through the superconducting material without any scattering events. The simplest way to achieve such an effect is to add and remove those conduction band electrons whose wavefunction has macroscopically large wavelength. Such a large wavelength no longer scatters because of the many orders of magnitude mismatch with respect to the inter-particle distances. This simplest model naturally suggests some sort of Bose-Einstein condensation, facilitating a variable particle number in the lowest-energy delocalized states. But despite intense efforts, there is no generally accepted theory yet that explains low- and high-temperature superconductivity phenomena in terms of Bose-Einstein condensation.

A main shortcoming of preceding Bose-Einstein condensation models is that they apply Bose-Einstein statistics only below the condensation temperature, while leaving open the question of applicable particle statistics above the condensation temperature. Taking any Bose-Einstein condensing particle population, these particles occupy quantum mechanical states also above the condensation temperature, and one must then clarify: how many particles per state are allowed above the condensation temperature, and what principle determines whether the incoherent particle population condenses partly or fully upon crossing below the condensation temperature? A meaningful discussion of these questions has been hindered by a historic ambiguity around particle statistics rules. Against this backdrop, the results of chapter 1 represent a major progress: our results clarify the exclusion principle's physical origin, and show that the applicability of Fermi-Dirac versus Bose-Einstein statistics is thermodynamically determined and therefore not any inherent property of a given particle.

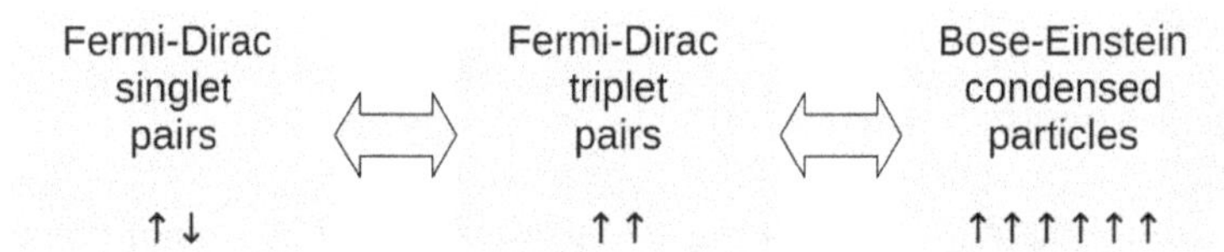

FIGURE 3.1.1. The proposed thermodynamic model: equilibrium transitions among the three particle populations are illustrated by the block arrows. The vertical arrows illustrate spin correlations.

Based on chapter 1 results, let us consider a particle dynamics model where the incoherent state is governed by Fermi-Dirac statistics, and the coherent state is governed by Bose-Einstein statistics. A thermodynamic balance is maintained between the coherent and incoherent particle populations. Within the incoherent Fermi-Dirac state, any quantum mechanical state is occupied by a particle pair in either spin-singlet or spin-triplet state[1]. Since a Bose-Einstein condensation into the same quantum mechanical state requires the alignment of particle spins, the direct thermodynamic balance is maintained between the Bose-Einstein condensed and spin-triplet paired particle populations. Figure 3.1.1 illustrates the thermodynamic balances among these three particle populations; evaluating these balances is the basis of our methodology.

Generally, the spin-singlet state is energetically favored, and we find it in virtually all materials. However, some rare materials, such as $K_2Cr_3As_3$, host a spin-triplet paired electron band **[20]**. When the parallel aligned spin-triplet pairs condense into a spin-aligned Bose-Einstein condensate, the scientific literature refers to it as a "spin-triplet superconductor".

In the following, the Bose-Einstein condensation temperature is calculated according to the above-outlined concepts. These calculations are based on well-known methods of statistical physics. While preceding Bose-Einstein condensation calculations lead to divergence in the two-dimensional case, our methodology is applicable to both three-dimensional and two-dimensional scenarios, as well as the anisotropic case that lies in-between these two topologies.

3.2. Quantum mechanical standing waves of enclosed particles

The quantum mechanical states of enclosed particles are standing waves, located in a square potential well that is defined by the enclosure. The particle wavefunctions reflect back from the enclosure boundary. This work considers systems where the interaction

[1]Occasionally, a quantum mechanical state can be also occupied by a single unpaired particle, but that is very rare in comparison to pair-wise occupancy.

among particles is negligible[2]; quantum mechanical standing waves are then simple harmonic waves. Examples where such a non-interacting harmonic wavefunction model applies: enclosed noble gases, or delocalized electrons within the interior of a metal. In the delocalized electron case, their electrostatic potential is counter-balanced by the positively charged lattice nuclei.

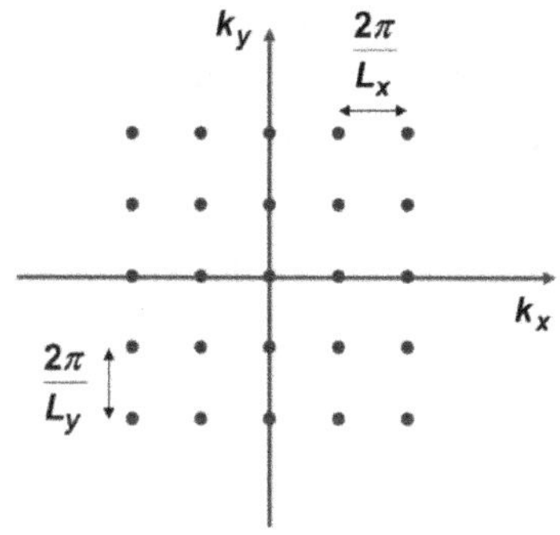

FIGURE 3.2.1. A k-space representation of delocalized electron states

Consider a cube shaped enclosure with dimensions L_x, L_y, L_z. A standing wave's wavenumbers may take on the values illustrated in figure 3.2.1, specifically:

$$k_x = l\frac{2\pi}{L_x},\ k_y = m\frac{2\pi}{L_y},\ k_z = n\frac{2\pi}{L_z},$$

where $l, m, n = 0, \pm 1, \pm 2, \ldots$.

When the particle movement is restricted to a 2-dimensional plane, the allowed wavenumbers are:

$$k_x = l\frac{2\pi}{L_x},\ k_y = m\frac{2\pi}{L_y},$$

with $l, m = 0, \pm 1, \pm 2, \ldots$.

Each gridpoint of figure 3.2.1 represents an allowed state. In the k-space of figure 3.2.1, there is only one state per every $\frac{(2\pi)^3}{L_x L_y L_z}$ volume element. Thus the k-space density of allowed states is:

$$\varrho_k = \frac{L_x L_y L_z}{(2\pi)^3}. \tag{3.2.1}$$

Similarly, the k-space density of 2-dimensional states is:

$$\sigma_k = \frac{L_x L_y}{(2\pi)^2}. \tag{3.2.2}$$

The kinetic energy of any given state is:

---

[2]When the interaction among particles is non-negligible, the allowed quantum mechanical states are solutions of the Gross–Pitaevskii equation. This equation incorporates interactions among particles through a pseudopotential, and its non-linearity arises from these interactions. The density of states formula then becomes more complex than equations 3.2.8 and 3.2.9. The rest of the article applies in the same way, but the appropriate density of states formulation must be used. In the context of superconductivity, the Gross–Pitaevskii equation has been shown to be relevant for sub-micron sized mesoscopic superconductors, whose size is comparable to the superconducting coherence length [**2**].

$$E = \frac{1}{2}mv^2 = \frac{p^2}{2m} = \frac{(\hbar k)^2}{2m}, \tag{3.2.3}$$

where $k^2 = k_x^2 + k_y^2 + k_z^2$ for the 3-dimensional case and $k^2 = k_x^2 + k_y^2$ for the 2-dimensional case.

We want to determine the number of delocalized particle states that are within a k_F threshold wavenumber. In the k-space representation, these states occupy a Fermi-sphere, with radius k_F. The number of delocalized states within k_F radius is:

$$N_{3d} = \varrho_k \frac{4\pi}{3} k_F^3 . \tag{3.2.4}$$

We denote the total density of particles by $n_{3d} = \frac{N_{3d}}{V}$. The relationship between the particle density and k_F is:

$$n_{3d} = \frac{k_F^3}{6\pi^2} . \tag{3.2.5}$$

In the case of a 2-dimensional plane, delocalized particles occupy a Fermi-disc, with radius k_F. We use the η_{2d} symbol to refer to the areal density, as it is dimensionally different from the volumetric density. The relationship between $\eta_{2d} = \frac{N_{2d}}{A}$ and k_F becomes:

$$\eta_{2d} = \frac{k_F^2}{4\pi} . \tag{3.2.6}$$

The energy of particles at k_F is called the Fermi energy:

$$E_F = \frac{(\hbar k_F)^2}{2m} . \tag{3.2.7}$$

Now we calculate the density of energy eigenstates as a function of particle energy. Using equation 3.2.3, we get $k = \frac{\sqrt{2mE}}{\hbar}$ and the following expression for $d_{3d}(E)$:

$$\begin{aligned} d_{3d}(E) \equiv \frac{dn_{3d}}{dE} = \frac{dn_{3d}}{dk}\frac{dk}{dE} = \frac{k^2}{2\pi^2}\frac{\sqrt{2m}}{\hbar}\frac{1}{2\sqrt{E}} = \\ = \left(\frac{\sqrt{m}}{\hbar}\right)^3 \frac{1}{\sqrt{2}\pi^2}\sqrt{E} . \end{aligned} \tag{3.2.8}$$

The density of three-dimensional particle states grows with the square root of particle energy.

An analogous calculation of $d_{2d}(E)$ yields:

$$\begin{aligned} d_{2d}(E) \equiv \frac{d\eta_{2d}}{dE} = \frac{d\eta_{2d}}{dk}\frac{dk}{dE} = \frac{k}{2\pi}\frac{\sqrt{2m}}{\hbar}\frac{1}{2\sqrt{E}} = \\ = \frac{m}{\hbar^2}\frac{1}{2\pi} . \end{aligned} \tag{3.2.9}$$

The above equation means that the density of particle states is constant in the two-dimensional case.

3.3. The entropy of a delocalized configuration

Suppose that we have N indistinguishable particles that are arranged among M available states. In this context, indistinguishability means that the exchange of any two particles does not impact any thermodynamic property, such as the number of available states or the energy levels of these states. How many configurations are available to distribute them? Distributing into M states is analogous to putting each particle into one of M boxes in a row; these boxes have $M-1$ walls between them. Essentially, we have a row of N particles and $M-1$ walls, and we want to count their possible configurations. We count up to $N+M-1$ elements, and each consecutive element can be either a particle type or a wall type.

The number of configurations is:

$$W = \frac{(N+M-1)!}{N! \cdot (M-1)!} \,. \tag{3.3.1}$$

The entropy S of a particle configuration is defined as:

$$S = k_b \ln W \,, \tag{3.3.2}$$

where $k_b$is the Boltzmann constant, and W is the number of available configurations.

According to equation 3.2.3, an electron's energy depends on the absolute value of its wavenumber k. However, each k state can be realized through various combinations of k_x, k_y, k_z wavenumbers: these possibilities are denoted by the M_k available states.

Let N_k be the number of particles with wavenumber k. In terms of thermodynamics, all N_k particles are indistinguishable: each one has the exact same kinetic energy. The number of configurations for all particles is given by:

$$W = \prod_k \frac{(N_k+M_k-1)!}{N_k! \cdot (M_k-1)!} \,. \tag{3.3.3}$$

We use Stirling's approximation for calculating the logarithms:

$$\ln n! \approx n \ln n - n \,,$$

which converges for large n. The total entropy then becomes:

$$S = k_b \ln W = k_b \sum_k \left[(N_k+M_k) \ln (N_k+M_k) - N_k \ln N_k - M_k \ln M_k \right] \,, \tag{3.3.4}$$

where we used the $N_k+M_k-1 \approx N_k+M_k$ approximation. Differentiating S with respect to a selected N_k variable yields:

$$\frac{\partial S}{\partial N_k} = k_b \left[\ln (N_k+M_k) + 1 - \ln N_k - 1 \right] \,.$$

The total kinetic energy is:

$$U=\sum_k E_k N_k\,, \tag{3.3.5}$$

where E_k is calculated from equation 3.2.3, and U remains constant for a closed system. In thermal equilibrium, the particles distribute themselves such that the total entropy is maximized. For a given number of particles, any variation around thermal equilibrium keeps the total number of particles constant:

$$N=\sum_k N_k\,. \tag{3.3.6}$$

Therefore, in order to find the thermodynamically favored equilibrium configuration, we must maximize S under the constraints of fixed $U$and N. This implies for each N_k the following condition:

$$\frac{\partial S}{\partial N_k}-k_b\beta\frac{\partial U}{\partial N_k}+k_b\beta\cdot\mu\frac{\partial N}{\partial N_k}=0\,, \tag{3.3.7}$$

where β and μ have been defined as optimization constants. For an adiabatic process, the above differential becomes $\frac{\partial U}{\partial N_k}=\mu\frac{\partial N}{\partial N_k}$, which implies that μ is in fact the chemical potential. Since a potential is defined with respect to a reference, to calculate μ we must know where the new particles are added from. Evaluating the above differentials yields:

$$\ln\left(N_k+M_k\right)-\ln N_k-\beta E_k+\beta\mu=0\,. \tag{3.3.8}$$

We solve the above equation for N_k:

$$N_k=\frac{1}{e^{\beta(E_k-\mu)}-1}M_k\,. \tag{3.3.9}$$

The average occupancy number of states with energy E_k is:

$$f\left(E_k\right)=\frac{N_k}{M_k}=\left(e^{\beta(E_k-\mu)}-1\right)^{-1}\,. \tag{3.3.10}$$

The above formula is known as the Bose-Einstein distribution. We are interested in finding the thermodynamic conditions that allow coherent particles' Bose-Einstein condensation, with the distribution given by equation 3.3.10.

We now evaluate the β parameter. When the number of particles is allowed to vary, the change in entropy is calculated using equation 3.3.7:

$$\begin{aligned} dS &= \sum_k \frac{\partial S}{\partial N_k}dN_k \\ &= k_b\beta\sum_k\left(\frac{\partial U}{\partial N_k}-\mu\frac{\partial N}{\partial N_k}\right)dN_k \\ &= k_b\beta\left(dU-\mu\cdot dN\right)\,. \end{aligned} \tag{3.3.11}$$

At the same time, dU is also given by the first law of thermodynamics:

$$dU=T\cdot dS-P\cdot dV+\mu\cdot dN\,,$$

where T is temperature, P is pressure, and μ is the chemical potential. Since there is no volume change within a fixed enclosure, the above equation leads to the following formulation of the entropy change:

$$dS = \frac{1}{T}\left(dU - \mu \cdot dN\right) . \tag{3.3.12}$$

By comparing equations 3.3.11 and 3.3.12, we find that:

$$\beta = \frac{1}{k_b T} . \tag{3.3.13}$$

3.4. The continuum limit of large distributions

Using 3.3.10, the total number of particles is:

$$N = \sum_k f\left(E_k\right) = \sum_k \left(e^{\beta(E_k-\mu)} - 1\right)^{-1} . \tag{3.4.1}$$

Recalling the density of k-space elements given by equation 3.2.1, in the continuum limit one may replace the summation by integrating over all possible wavenumbers:

$$N = \frac{V}{(2\pi)^3}\int \left(e^{\beta(E_k-\mu)} - 1\right)^{-1} d^3k ,$$

which gives the following particle density:

$$n = \frac{1}{(2\pi)^3}\int \left(e^{\beta(E_k-\mu)} - 1\right)^{-1} d^3k .$$

Replacing the wavenumber integration by integrating over particle energy values, we obtain the following expression for the density of delocalized particles, for the 3-dimensional and 2-dimensional cases respectively:

$$n = \int_0^\infty \left(e^{\beta(E-\mu)} - 1\right)^{-1} d_{3d}\left(E\right) dE , \tag{3.4.2}$$

$$\eta = \int_0^\infty \left(e^{\beta(E-\mu)} - 1\right)^{-1} d_{2d}\left(E\right) dE ,$$

where the $d_{3d}\left(E\right)$ parameter of equation 3.2.8 gives the density of particle states at a given energy.

Let us consider the role of chemical potential μ. In our context, the Fermi-sea is the source of electrons during a dN change in their total number. If a stable Fermi-sea and Bose-Einstein condensate co-exist, the $\mu = 0$ condition represents their balance[3]. Considering the finite number of available particles, the density of Bose-Einstein condensed particles can be summarized by the following two rules:

[3]The $\mu \neq 0$ condition means that energy is gained by a particle moving from the Fermi-sea to tha Bose-Einstein condensate, or vice versa. Therefore, under the $\mu \neq 0$ condition, a thermodynamic balance between a Fermi-sea and Bose-Einstein condensate would require an electrostatic potential difference to counter-balance the chemical potential difference. Within any given material, the Fermi-sea and Bose-Einstein condensate occupy the same region; i.e. there cannot be any electrostatic potential difference between them. This leads to the conclusion that the $\mu = 0$ condition is a pre-requisite for the co-existing Fermi-sea and Bose-Einstein condensate being in thermodynamic balance within any given material.

(1) If equation 3.4.2 can be solved in $\mu = 0$ case, the spin-triplet Fermi-sea particles flow into the Bose-Einstein condensate till the $\mu = 0$ condition is established; the Fermi-sea and Bose-Einstein condensate co-exist at the given temperature.
(2) If equation 3.4.2 cannot be solved for $\mu = 0$, all spin-triplet Fermi-sea particles flow into the Bose-Einstein condensate when it becomes thermodynamically favorable. A thermodynamic precondition is that the pre-condensation mean energy in the Dirac-Fermi sea must be at least as high as the mean energy of the Bose-Einstein condensate.

If the $\mu = 0$ condition can be fulfilled, we arrive at the following expression for the density of particles:

$$n = \int_0^\infty \left(e^{\beta E} - 1\right)^{-1} d_{3d}(E)\, dE\,. \tag{3.4.3}$$

In the above expression, the $\left(e^{\beta E} - 1\right)^{-1}$ term diverges at the $E = 0$ energy value. This divergence expresses that having an infinite reservoir of Fermi-sea particles pulls an infinite number of them into the Bose-Einstein ground state. It is exactly the dynamics that characterizes Bose-Einstein condensates: particles are continuously condensed into the ground state as long as there is an available supply. Despite this ground state divergence, as we shall see below, the overall integral still converges to a finite value.

3.5. Bose-Einstein condensation

The phase transition that we derive in this section is therefore unique in the sense that it is driven by delocalized wavefunction interactions, and not by local interactions between nearby particles. In the following, we calculate Bose-Einstein condensation dynamics for three topological scenarios.

3.5.1. 3-dimensional isotropic case. This scenario is relevant for non-interacting gases. Also, elementary metal based superconductors fall into this category, as evidenced by their isotropic diamagnetism.

Let us re-write equation 3.4.3 in terms of the dimensionless $x := \beta E$ variable:

$$n_{BEC} = \left(\frac{\sqrt{mk_bT}}{\hbar}\right)^3 \frac{1}{\sqrt{2}\pi^2} \int_0^\infty \frac{e^{-x}}{1 - e^{-x}} \sqrt{x}\, dx\,. \tag{3.5.1}$$

To evaluate the above integral, we use the Maclaurin series expansion of $\frac{1}{1-e^{-x}}$:

$$\frac{e^{-x}}{1 - e^{-x}} = e^{-x}\left(1 + e^{-x} + e^{-2x} + \ldots\right) = \sum_{a=1}^{\infty} e^{-ax}\,.$$

For each a value, we evaluate the integral expression by making the $y := ax$ variable change:

$$\int_0^\infty e^{-ax}\sqrt{x}\,dx = a^{-\frac{3}{2}}\int_0^\infty e^{-y}\sqrt{y}\,dy = a^{-\frac{3}{2}}\Gamma\left(\frac{3}{2}\right),$$

where Γ is the so-called gamma function, with $\Gamma\left(\frac{3}{2}\right) = \frac{\sqrt{\pi}}{2}$. Substituting the above result into equation 3.5.1, we obtain:

$$n_{BEC} = \left(\frac{mk_bT}{2\pi\hbar^2}\right)^{\frac{3}{2}}\sum_{a=1}^{\infty}a^{-\frac{3}{2}}. \tag{3.5.2}$$

In the above equation, the summation result is given by the Riemann zeta function:

$$\sum_{a=1}^{\infty}a^{-\frac{3}{2}} = \zeta\left(\frac{3}{2}\right) \approx 2.612\,.$$

Equation 3.5.2 expresses the relationship between the Bose-Einstein condensed particle density n_{BEC} and temperature. As we obtained equation 3.5.2 for the case of $\mu = 0$, the temperature appearing in this equation is in fact the limiting temperature of Bose-Einstein condensation that allows a co-existence between the Fermi-sea and Bose-Einstein condensate particle states:

$$\left[n_{BEC} = \left(\frac{mk_bT_{BEC}}{2\pi\hbar^2}\right)^{\frac{3}{2}}\zeta\left(\frac{3}{2}\right)\right]_{3d}, \tag{3.5.3}$$

where the n_{BEC} density of Bose-Einstein condensed particles appears abruptly at the phase-change temperature, i.e. n_{BEC}=0 for $T > T_{BEC}$.

Finally, we rearrange the above equation to solve for T_{BEC}:

$$\left[T_{BEC} = \frac{2\pi\hbar^2}{mk_b}\left(\frac{n_{BEC}}{\zeta\left(\frac{3}{2}\right)}\right)^{\frac{2}{3}}\right]_{3d}. \tag{3.5.4}$$

Equation 3.5.4 is an important result: it defines the limiting temperature of an isotropic Bose-Einstein condensation. This equation has been known in the scientific literature; it was first derived by Fritz London in 1938 [**3**], and is nowadays further explained in textbooks [**4**]. However, equation 3.5.4 was previously derived by assuming the Bose-Einstein condensation of all available particles, and without investigating the chemical potential's role. Here we see that fulfilling the $\mu = 0$ condition means that the Fermi-Dirac and Bose-Einstein particle populations co-exist at any $T > 0$ temperature. Only in the $T \to 0$ limit do all Fermi-Dirac particles Bose-Einstein condense.

Let us denote by $n_{singlet}$ and $n_{triplet}$ the density of spin-singlet and spin-triplet Fermi-Dirac particles, evaluated just before the Bose-Einstein condensation. As it is the spin-triplet particles that are capable of Bose-Einstein condensing, it follows that $n_{triplet} \geq n_{BEC}(T_{BEC})$.

In elementary metals, $n_{BEC}(T_{BEC})$ is a tiny fraction of the total density of delocalized electrons. Let us consider lead, as a representative elementary metal superconductor. Based on lead crystal's 0.5 nm unit cell dimension and the +2 valence state of lead sites, we estimate n=(0.4 nm)$^{-3}$ total delocalized electron density in lead metal. Applying T_c=7.2 K of lead metal to equation 3.5.3, we obtain that $n_{BEC}(T_{BEC})/n$ is only 0.05%. Lead material has $\lambda_L(T \to 0)$=52 nm magnetic field penetration depth and $m_{eff} = 1.9m_e$

effective electron mass; we can use the $\frac{1}{\lambda_L} = \sqrt{n_s \frac{\mu_0 e^2}{m_{eff}}}$ London equation, which was derived in chapter 2, to find the n_s density of superconducting electrons at $T \to 0$. Carrying out this calculation yields $n_s(T \to 0)$=(0.37 nm)$^{-3}$ superconducting electron density. This data demonstrates that $n_{BEC}(T_{BEC}) \ll n$ and $n_{BEC}(T \to 0) \approx n$ in elementary metals. Determining $n_{BEC}(T_{BEC})$ thus becomes the basis for estimating T_{BEC} in isotropic materials.

In hydrated or deuterated metals, the $n_{BEC}(T_{BEC})$ ratio can reach an order of magnitude higher value than in elementary metals, even at ambient pressure. The authors of [**5**] achieved 60 K ambient pressure superconductivity in rapidly cooled deuterated palladium (PdD); this is the highest superconducting temperature so far achieved in a binary isotropic material. The electronic structure of hydrated palladium has been extensively studied, and these studies conclude that the palladium-hydrogen bond has mainly covalent character. The +2 valence of metallic palladium therefore turns into approximately +1 valence in PdD. Based on PdD crystal's 0.414 nm unit cell dimension and +1 valence, we estimate n=(0.414 nm)$^{-3}$ total delocalized electron density. Applying T_c=60 K of PdD metal to equation 3.5.3, we obtain that $n_{BEC}(T_{BEC})/n$ is 0.5% in PdD. The different electron excitation process in elementary versus hydrated metals is illustrated by their opposite nuclear mass dependence: heavier lead isotopes yield lower T_c, while heavier hydrogen isotopes yield higher T_c in hydrated palladium. Specifically the authors of [**5**] achieved 60 K superconductivity with PdD, but only 54 K superconductivity with PdH. The involved catalytic processes shall be examined in chapter 5.

The Buckminsterfullerene based Rb_3C_{60} material is an example of even higher $n_{BEC}(T_{BEC})/n$ fraction. It comprises 1.45 nm spacing between the +1 valence Rb sites, and thus we estimate n=(1.45 nm)$^{-3}$ total delocalized electron density. Applying T_c=29 K of Rb_3C_{60} metal to equation 3.5.3, we obtain that $n_{BEC}(T_{BEC})/n$ is 3% in Rb_3C_{60}. Table 1 summarizes the two orders of magnitude $n_{BEC}(T_{BEC})/n$ variation across these reviewed materials. Such large variation in the $n_{BEC}(T_{BEC})/n$ ratio demonstrates the existence of catalytic processes that determine the rate of spin-triplet pair production within the Fermi-Dirac sea, or even the production of unpaired electrons in the vicinity of the Fermi energy level.

Superconductor	T_{BEC}	$n_{BEC}(T_{BEC})/n$
Pb	7.2 K	0.05%
PdD (slowly cooled)	8.2 K	0.07%
PdD (rapidly cooled)	60 K	0.5%
Rb_3C_{60}	29 K	3%

TABLE 1. The fraction of delocalized electrons that Bose-Einstein condense at isotropic materials' superconducting transition temperature.

We derived equation 3.5.4 for the $T = T_{BEC}$ case, and the next step is to understand what happens in the $T < T_{BEC}$ case. We discussed in section 3.4 that the $\left(e^{\beta E} - 1\right)^{-1}$ term diverges at the $E = 0$ energy value, which means that a significant fraction of electrons condenses into the quantum mechanical ground state. Therefore, in the $T < T_{BEC}$ case we must treat separately those N_0 electrons that occupy the E_0 ground state.

We therefore formulate equation 3.4.1 as follows, keeping in mind the $\mu = 0$ condition:

(3.5.5)
$$N = N_0 + \sum_{k \neq 0} \left(e^{\beta E_k} - 1\right)^{-1} .$$

In the continuum limit, the corresponding electron density becomes:

(3.5.6)
$$\left[n_{BEC} = n_0 + \left(\frac{\sqrt{mk_bT}}{\hbar}\right)^3 \frac{1}{\sqrt{2}\pi^2} \int_0^\infty \frac{e^{-x}}{1 - e^{-x}} \sqrt{x}\, dx = n_0 + \left(\frac{mk_bT}{2\pi\hbar^2}\right)^{\frac{3}{2}} \zeta\left(\frac{3}{2}\right) \right]_{3d} .$$

Dividing by n_{BEC} and substituting equation 3.5.4, we obtain:

(3.5.7)
$$\left[\frac{n_0}{n_{BEC}} = 1 - \left(\frac{T}{T_{BEC}}\right)^{\frac{3}{2}} \right]_{3d} .$$

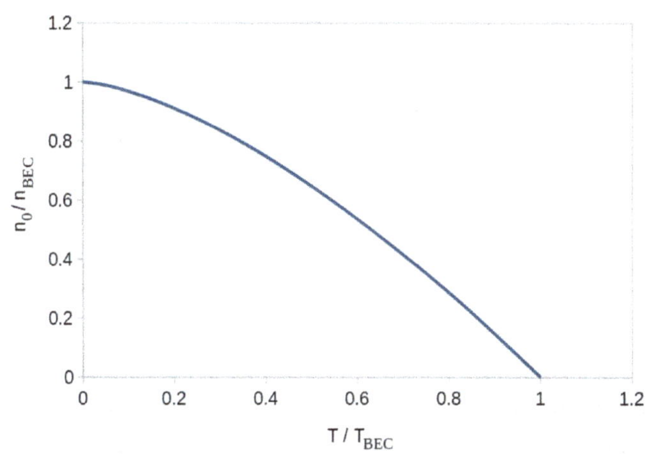

FIGURE 3.5.1. The $\frac{n_0}{n_{BEC}}$ ratio of Bose-Einstein condensed electrons that occupy the ground state in isotropic electron condensates, as a function of $\frac{T}{T_{BEC}}$.

Equation 3.5.7 defines the ratio of Bose-Einstein condensed electrons that occupy the quantum mechanical ground state, and it is graphically illustrated in figure 3.5.1. It seen that nearly all Bose-Einstein condensed electrons occupy the ground state at $T \to 0$. The percentage Bose-Einstein condensed electrons that occupy the ground state gradually goes to zero in the $T \to T_{BEC}$ temperature limit. This result explains the experimental observation of $\lambda_L (T \to T_c) \to \infty$ in isotropic superconducting materials; the $\frac{1}{\lambda_L} = \sqrt{n_s \frac{\mu_0 e^2}{m_{eff}}}$ London equation tells us that the infinite λ_L value corresponds to zero density of electrons in the superconducting ground state at $T \to T_c$ temperature, exactly as predicted by equation 3.5.7.

3.5.2. Stacked 2-dimensional case. In this section, we describe Bose-Einstein condensation for the purely 2-dimensional case. Let us firstly calculate the 2-dimensional case for $\mu = 0$, starting from the 2-dimensional analogue of equation 3.4.3:

$$\eta_{BEC} = \int_0^\infty \left(e^{\beta E} - 1\right)^{-1} d_{2d}(E)\, dE\,, \tag{3.5.8}$$

where the $d_{2d}(E)$ parameter of equation 3.2.9 gives the density of electron states at a given energy.

As before, we re-write this equation in terms of the dimensionless $x := \beta E$ variable:

$$\eta_{BEC} = \frac{m}{2\pi\hbar^2} k_b T \int_0^\infty \frac{e^{-x}}{1 - e^{-x}}\, dx\,. \tag{3.5.9}$$

To evaluate the above integral, we again use the $\frac{e^{-x}}{1-e^{-x}} = \sum_{a=1}^{\infty} e^{-ax}$ identity. For each a value, the integral expression yields:

$$\int_0^\infty e^{-ax}\, dx = \frac{1}{a}\,.$$

The integral expression of equation 3.5.9 thereby evaluates to:

$$\sum_{a=1}^{\infty} a^{-1} = \zeta(1)\,.$$

The above summation diverges as the Riemann zeta function $\zeta(x)$ grows to infinity in the $x \to 1$ limit. The divergence of $\zeta(1)$ means that the density of Bose-Einstein condensate grows as long as there is a supply of electrons, i.e. there is no thermodynamically favorable co-existence between a 2-dimensional Bose-Einstein condensate and Fermi-Dirac sea. The 2-dimensional case is therefore simpler than the 3-dimensional one: if Bose-Einstein condensation becomes thermodynamically favored, all delocalized electrons Bose-Einstein condense via the mediation of spin-triplet electrons, implying $\eta_{BEC} = \eta$ already at the superconducting transition temperature.

We calculate T_{BEC} from the thermodynamic precondition that the pre-condensation mean energy in the Dirac-Fermi sea must be at least as high as the mean energy of the Bose-Einstein condensate. At any given temperature, the Bose-Einstein condensate's mean energy is $E_{mean} = k_b T$.

There are two electrons per energy eigenstate in the Dirac-Fermi sea. Recalling from equation 3.2.9 that the energy eigenstates are evenly spaced in a 2-dimensional metal, the mean energy of the Dirac-Fermi sea is $\frac{1}{2}$ of its Fermi energy level. It follows that $\eta_{BEC} = \eta$ is given by the following formula:

$$\eta = \int_0^{2E_{mean}} 2d_{2d}\, dE = 2d_{2d} \cdot (2E_{mean}) = \frac{2m}{\pi\hbar^2} k_b T_{BEC}\,. \tag{3.5.10}$$

We rearrange the above equation to solve for T_{BEC}:

$$T_{BEC} = \frac{\pi\hbar^2}{2mk_b}\eta\,. \tag{3.5.11}$$

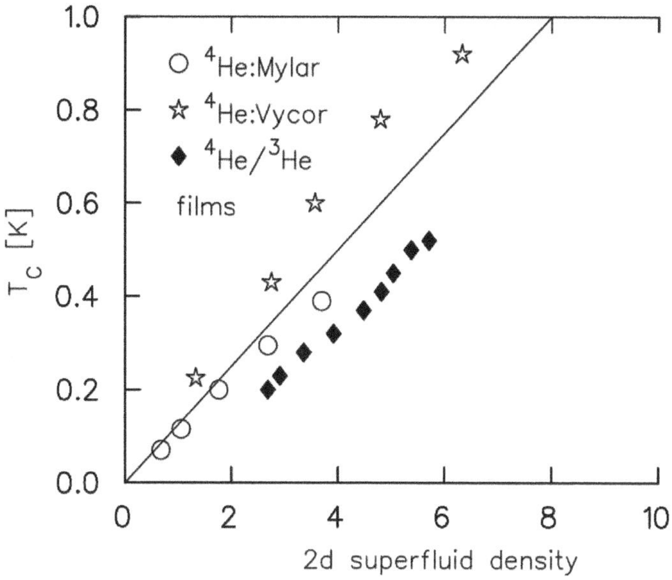

FIGURE 3.5.2. The $T_{BEC} \sim \eta$ relationship in nearly 2-dimensional superfluids, reproduced from **[6]**

Equation 3.5.11 implies a linear $T_{BEC} \sim \eta$ relationship. Regarding experiments, nearly 2-dimensional Bose-Einstein condensation has been studied in the context of thin superfluid films **[6]**. As shown in figure 3.5.2, the anticipated $T_{BEC} \sim \eta$ approximate relationship is indeed experimentally observed.

While a purely 2-dimensional case may be a mathematical abstraction, strongly anisotropic materials approach this limit, and equation 3.5.11 becomes a good approximation for strongly anisotropic materials.

3.5.3. Generalized anisotropic case. Next, we consider a generalized case in-between the 3-dimensional isotropic and the stacked 2-dimensional topologies. In the context of superconductors, the anisotropic topology means that conduction band electrons have 3-dimensional delocalization, but their superconducting diamagnetism has significant anisotropy. Practically all high-temperature superconductors falls into this category. We model this class of materials by interpolating between the 3-dimensional and 2-dimensional topologies.

When the particle states are not purely 2-dimensional, a solution for μ=0 shall exist. Revisiting equation 3.5.9, we firstly carry on with the 2-dimensional calculation, using the symbolic $\zeta(1)$ notation.

Let d_{il} denote the inter-layer spacing between the crystal planes that more strongly conduct electrons. In order to compare 3-dimensional versus 2-dimensional results using dimensionally matching equations, we shall use the $n_{BEC} := \eta d_{il}^{-1}$ density notation for the 2-dimensional computation. Using equation 3.5.9, we can now express the Bose-Einstein condensed electron density in terms of its limiting temperature:

$$\left[n_{BEC} = \zeta(1)\,\frac{mk_b T_{BEC}}{2\pi\hbar^2 d_{il}}\right]_{2d}. \tag{3.5.12}$$

We rearrange the above equation to solve for T_{BEC}:

$$\left[T_{BEC}=\frac{2\pi\hbar^2}{mk_b}d_{il}\frac{n_{BEC}}{\zeta\left(1\right)}\right]_{2d}. \tag{3.5.13}$$

For the generalized anisotropic case, the dimensionality of electron conduction is somewhere between 2-dimensional and 3-dimensional. By comparing equation 3.5.13 against 3.5.4, it is possible to formulate the generalized T_{BEC} formula, and there is only one dimensionally correct way of doing it[4]:

$$\left[T_{BEC}=\frac{2\pi\hbar^2}{mk_b}\left(d_{il}\right)^{(3D-2)}\left(\frac{n_{BEC}}{\zeta\left(D^{-1}\right)}\right)^D\right]_{anisotropic}, \tag{3.5.14}$$

where D is the dimensionality parameter. T_{BEC} is well-defined in the $2/3 \geq D > 2/2$ range, and isotropic materials have the $D = 2/3$ value. Experimentally, the value of D can be inferred via London penetration depth measurements along the various axes of the crystal structure. As before, it follows from the $\mu = 0$ condition that the Fermi-Dirac and Bose-Einstein condensed phases co-exist at any $T > 0$ temperature.

It is possible to calculate the $\frac{n_0}{n_{BEC}}$ ratio of ground state occupying electrons; the calculation procedure is analogous to how equation 3.5.7 was obtained in the isotropic case. For the nearly two-dimensional case, where $D \to 2/2$, we obtain the following linear $\frac{n_0}{n_{BEC}}$ ratio:

$$\left[\frac{n_0}{n_{BEC}}=1-\frac{T}{T_{BEC}}\right]_{2d} \tag{3.5.15}$$

Therefore, even in the two-dimensional limit, the percentage Bose-Einstein condensed electrons that occupy the ground state gradually goes to zero in the $T \to T_{BEC}$ temperature limit. This result explains the experimental observation of $\lambda_L\left(T \to T_c\right) \to \infty$ in all superconducting materials. The isotropic case is represented by elementary metals and the $\frac{n_0}{n_{BEC}}$ ratio is given by equation 3.5.7, while the nearly two-dimensional case is represented by cuprate type superconductors and the $\frac{n_0}{n_{BEC}}$ ratio is given by equation 3.5.15.

[4]Specifically, the $\left(\frac{n_{BEC}}{\zeta(D^{-1})}\right)^D$ term is the generalization of the 3-dimensional $\left(\frac{n_{BEC}}{\zeta(\frac{3}{2})}\right)^{\frac{2}{3}}$ and 2-dimensional $\left(\frac{n_{BEC}}{\zeta(\frac{2}{2})}\right)^{\frac{2}{2}}$ terms. The $\frac{2\pi\hbar^2}{mk_b}$ factor remains the same for any topology. In order for the formula to yield temperature and to match with the generalized particle density definition, the d_{il}^{3D-2} term must be present as well. I.e. the dimensionally generalized particle density is $n_{BEC}^D \cdot d_{il}^{3D-2}$, where n_{BEC} is the 3-dimensional particle density and d_{il} is the inter-layer distance of the crystal structure.

3.6. Experimental validation

3.6.1. Bose-Einstein condensation from spin-triplet paired electron state. $K_2Cr_3As_3$ is an exemplary material that is thought to contain an electron band comprising delocalized electrons with parallel spin correlation [**20**]. Such electron pairs are also being referred to as a "spin-triplet" state.

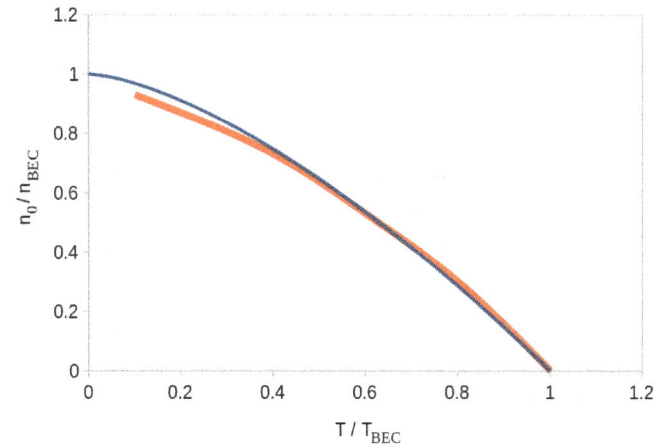

FIGURE 3.6.1. The $\frac{n_0}{n_{BEC}}$ ratio as a function of $\frac{T}{T_{BEC}}$ in $K_2Cr_3As_3$. The red curve is the experimental data from [**19**], and the blue curve corresponds to equation 3.5.7.

The existence of a spin-triplet comprising Fermi-Dirac sea contradicts the historically formulated electron statistics postulates. On the other hand, a spin-triplet comprising Fermi-Dirac sea is perfectly compatible with the results of chapter 1, which proves the pair-wise Pauli exclusion limit of incoherent electron states, regardless of their parallel or anti-parallel spin correlation.

Electronic structure studies on $K_2Cr_3As_3$ show that it comprises three distinct delocalized electron bands: two are nearly 1-dimensional, and one is nearly isotropic [**19, 20**]. Superconductivity is associated with the phase transition of the isotropic electron band, and this nearly isotropic band is thought to host spin-triplet electron pairs.

Despite its relatively low superconducting temperature of T_c=6.2 K, $K_2Cr_3As_3$ maintains its superconducting state even under >20 T magnetic field at $T \to 0$ temperature. While lead metal has a similar superconducting temperature of T_c=7.2 K, lead looses its superconductivity already under 0.08 T magnetic field at the same $T \to 0$ temperature. Because of these properties, $K_2Cr_3As_3$ is being classified as an "unconventional" superconductor.

Considering that the superconductivity-related topology is nearly isotropic in $K_2Cr_3As_3$, we compare its experimental data against the equations of section 3.5.1. Its superconducting electron density can be characterized by magnetic field penetration depth measurements. As shown in figure 3.6.1, the experimental superconducting electron density data closely matches equation 3.5.7.

The matching data of figure 3.6.1 in fact requires that n_{BEC} remains constant across the charted temperature range. This implies that all (or most) spin-triplet state electrons Bose-Einstein condense at the $T_c = T_{BEC}$ transition temperature, which validates the thermodynamic equilibrium model sketched in figure 3.1.1.

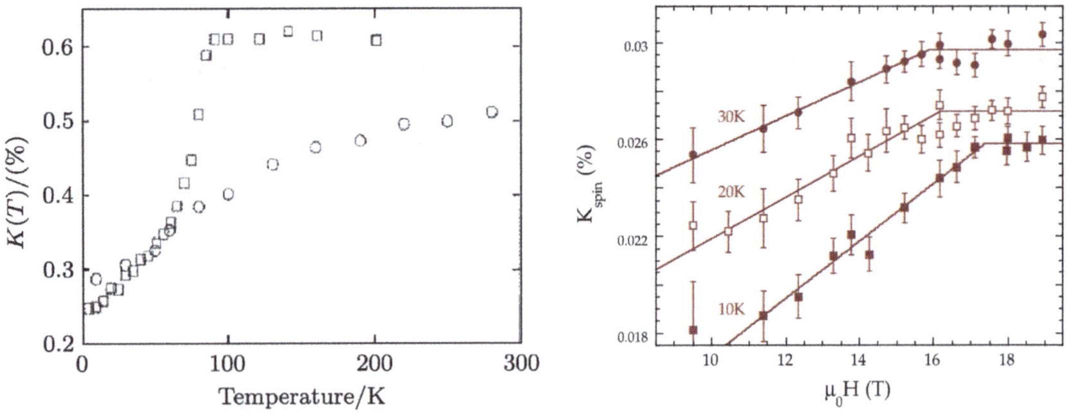

FIGURE 3.6.2. The signature of un-paired electron condensation is the Knight shift value drop in $YBa_2Cu_3O_{7-\delta}$, upon transition into superconducting state. Left: the Knight shift versus temperature, reproduced from **[10]**. The square symbols represent optimally doped $YBa_2Cu_3O_{6.9}$ with T_c=90 K, the round symbols represent under-doped $YBa_2Cu_3O_{6.7}$. Right: the Knight shift versus magnetic field strength for under-doped $YBa_2Cu_3O_{6.7}$, reproduced from **[11]**. The measurements were made at the various indicated temperatures, the Knight shift drop starts at B_c for the given temperature.

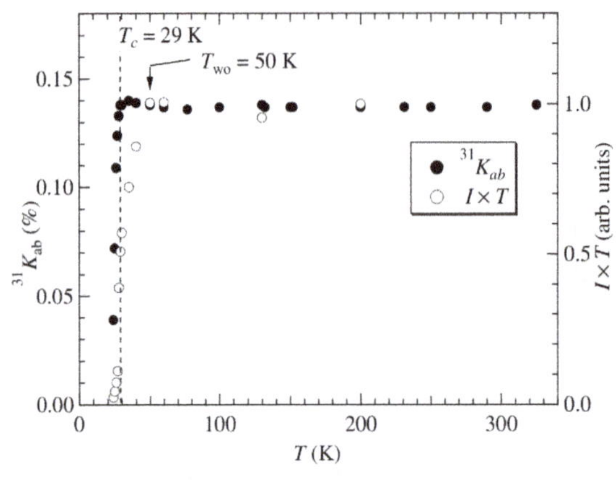

FIGURE 3.6.3. The signature of unpaired electron condensation is the Knight shift value drop in $Ba_{0.5}Sr_{0.5}Fe_2\,(As_{0.6}P_{0.4})_2$, upon transition into superconducting state. The Knight shift was measured by ^{31}P NMR method. Reproduced from **[12]**.

3.6.2. The transition of unpaired Fermi-Dirac electrons into the Bose-Einstein condensate. Although unpaired electron states are rare, the density of unpaired delocalized electrons can be measured via the NMR Knight shift which they produce. High-temperature superconductors' Knight shift value starts dropping exactly below their superconducting transition point. This phenomenon is illustrated in figure 3.6.2, taking the example of cuprate-type $YBa_2Cu_3O_{7-\delta}$ superconductor. In the case of optimally doped $YBa_2Cu_3O_{6.9}$, its Knight shift remains constant as the temperature is lowered towards T_c, and then starts sharply decreasing in the $T < T_c$ region. In the case of under-doped $YBa_2Cu_3O_{6.7}$, this phenomenon is much weaker, but still observable under variable applied magnetic field: the Knight shift remains constant as the magnetic field strength is lowered towards the critical magnetic field strength B_c, and then starts decreasing in the $B < B_c$ region. Analogous Knight shift charts are observed for other high temperature superconductors as well: figure 3.6.3 shows the example of iron-based $Ba_{0.5}Sr_{0.5}Fe_2\,(As_{0.6}P_{0.4})_2$ superconductor. This experimental phenomenon demonstrates that unpaired delocalized electrons are gradually depleted when the superconducting phase is present, as they condense one by one into the Bose-Einstein condensate. In other words, individual electrons may transition to and from the superconducting electron phase.

3.6.3. Isotropic superconductors. Equation 3.5.4 implies $T_c \sim n_{BEC}^{\frac{2}{3}}$ scaling for a given isotropic material, with all else being equal. This isotropic superconductor case is nicely illustrated by boron-doped diamond: it is a binary structured superconductor, where the density of conduction band electrons is proportional to the boron doping density. As illustrated in figure 3.6.4, the T_c of boron-doped diamond follows the anticipated $T_c \sim n_s^{\frac{2}{3}}$ relationship.

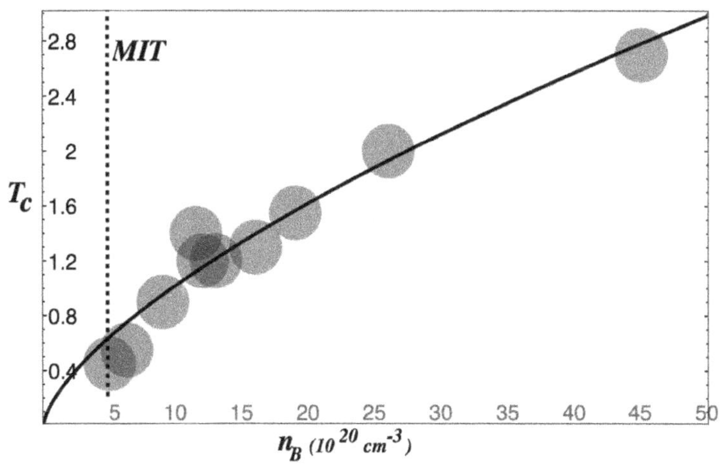

FIGURE 3.6.4. The T_c (K) of boron-doped diamond, as a function of boron concentration. The data is taken from [7], disc sizes correspond to the error bars. The solid line represents $\sim n^{\frac{2}{3}}$ scaling. The dashed line indicates metal-insulator transition.

3.6.4. Superconductors comprising stacked 2-dimensional topology.

3.6.4.1. *Optimally doped superconductors.* Nearly all high-temperature superconductors comprise stacked conducting planes, and their delocalized electron concentration is defined by the doping concentration. Their nearly 2-dimensional anisotropy can be characterized by the $\lambda_L\,(T \to 0)$ magnetic field penetration depth's anisotropy. We firstly consider "optimally doped" superconductors, where T_c is maximized. In optimally doped cuprate materials, which represent the most studied class of high-temperature superconductors, $\lambda_L\,(T \to 0)$ is 10-30 times larger in perpendicular direction to the CuO_2 planes than in parallel direction with respect to the CuO_2 planes. These cuprate materials can be therefore characterized as strongly 2-dimensional materials, in which superconducting electrons move mainly along the CuO_2 planes. We thus anticipate T_{BEC} to be given by equation 3.5.11.

In a remarkable work, the authors of [8] demonstrate a rather precise $T_c = \kappa x^{-2} N^{\frac{2}{3}}$ relationship for high-temperature superconductors, where x is the spacing between the doping sites that donate delocalized electrons, N is the number of CuO_2 planes within the unit cell, and $\kappa = 6.96 \cdot 10^{16}$ m^{2}K is a phenomenologically determined constant.

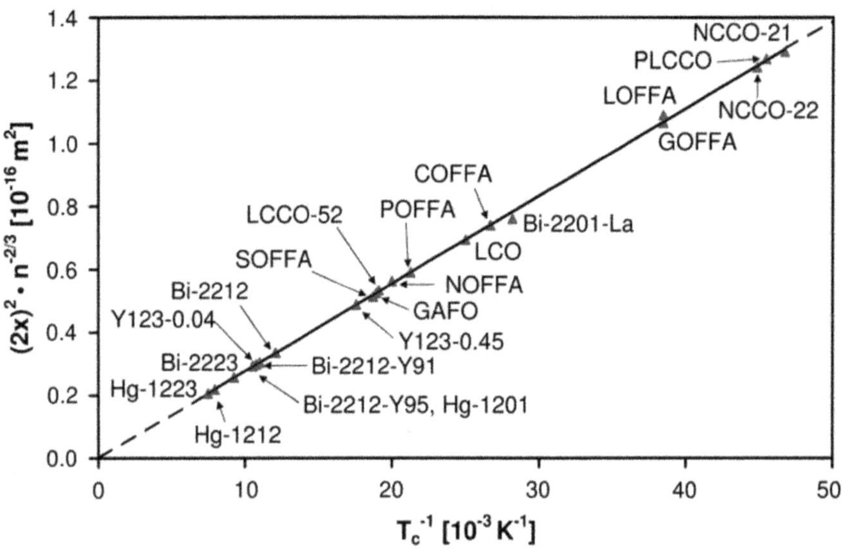

FIGURE 3.6.5. The Roeser-Huber relationship in iron-based and cuprate-based high-temperature superconductors, reproduced from [**6**]

This relationship is known as the Roeser-Huber formula, and it is illustrated in figure 3.6.5. Despite the Roeser-Huber formula's accuracy, so far it gained little attention among superconductivity researchers because it does not match with preceding superconductivity theories.

Comparing the Roeser-Huber formula against equation 3.5.11, we recognize that $x^{-2} \equiv \eta$ and $\frac{\pi\hbar^2}{2mk_b} \approx 2\kappa$. In most cuprate superconductors one finds N=3, and occasionally N=2 or N=4. The dimensionless $N^{\frac{2}{3}}$ parameter thus usually evaluates to $N^{\frac{2}{3}} \approx 2$. I.e. $\frac{\pi\hbar^2}{2mk_b} \approx N^{\frac{2}{3}}\kappa$, and it is clearly seen that the Roeser-Huber formula is in fact equation 3.5.11. The match between equation 3.5.11 and experimental data is therefore remarkable, and validates our theory.

3.6.4.2. *Underdoped superconductors.* The under-doped regime of high-temperature superconductors is characterized by a gradually growing delocalized electron concentration, as a function of the doping parameter. The concentration of delocalized electrons is approximately zero near the insulator-metal transition point, and gradually grows with increasing doping levels. Equation 3.5.11 indicates that the concentration of delocalized electrons must be the T_{BEC} limiting factor in this regime.

Around 1990, Y. Uemura found the $\lambda_L^{-2}(T \to 0) \sim T_c$ relationship, which universally holds for the underdoped regime of cuprate superconductors, and has been very unexpected by superconductivity theories at the time of its discovery. This relation is shown in figure 3.6.6. The data of figure 3.6.6 was obtained via muon spin relaxation measurement, where the muon spin relaxation rate σ is proportional to λ_L^{-2}. The $\frac{1}{\lambda_L} = \sqrt{n_s \frac{\mu_0 e^2}{m_{eff}}}$ London equation relates the $\lambda_L^{-2}(T \to 0)$ measurements to the superconducting electron density. For nearly 2-dimensional materials we can express the superconducting electron density using the previously discussed $n_s = \eta d_{il}^{-1}$ formula, where η is the areal density of Bose-Einstein condensed electrons, and d_{il} is the inter-layer distance of superconducting planes. At $T \to 0$, the Bose-Einstein condensed electrons all occupy the ground state, thus contributing to the superconducting current. Therefore the $\lambda_L^{-2}(T \to 0) \sim T_c$ observation leads to the $T_c \sim \eta_{BEC}$ relationship as long as m_{eff} and d_{il} remain constant. The obtained $T_c \sim \eta_{BEC}$ relationship matches with equation 3.5.11: i.e. we observe that Uemura's linear $\lambda_L^{-2}(T \to 0) \sim T_c$ relationship is a second validation of equation 3.5.11.

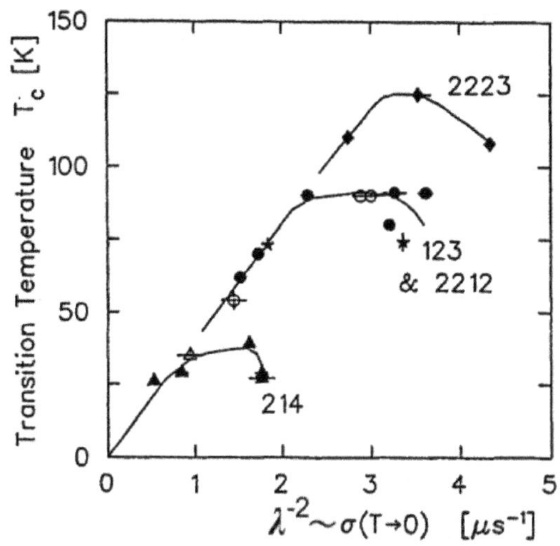

FIGURE 3.6.6. The $T_c \sim \lambda_L^{-2}(0)$ relationship, which is universally valid in the underdoped region of cuprates. The labels indicate various cuprate materials. Reproduced from [**9**].

At the optimal doping point, T_c deviates from the linear Uemura relationship observed in the underdoped regime, implying that m_{eff} or d_{il} become significantly impacted by the doping concentration. On the other hand, the Roeser-Huber formula is valid at the optimal doping point. From this point of view, the Uemura and Roeser-Huber relationships can be seen as complementary.

3.6.4.3. *Overdoped superconductors.* While the synthesis of optimally doped superconductors is relatively straightforward, further increases of doping concentration are experimentally more difficult to achieve. The synthesis of overdoped superconductors generally requires harsh oxidation methods, such as ozone or fluorine treatment.

For cuprate type superconductors, the highest T_c is generally observed near $p \approx 0.15$ doping concentration, where p measures the excess charge density with respect to the Cu concentration. The p>0.15 regime is the "overdoped region"; it is characterized by $T_c = T_{BEC}$ gradually descending to zero, and superconductivity completely disappears around $p \approx 0.25$. The corresponding superconducting phase diagram is illustrated in figure 3.6.7. Between 1990 and 2020, this phase diagram was thought to be universally valid for cuprate type superconductors. An analogous phase diagram was identified for iron-based superconductors as well.

Preceding high-temperature superconductivity theories were formulated around rationalizing the emergence of high-temperature superconductivity from a non-Fermi liquid ground state, motivated by the separation of these regions in the phase diagram of figure 3.6.7.

Through ingenious experiments, the authors of [**13**] and [**14**] demonstrate that the gradual disappearance of superconductivity in the over-doped regime is just the side-effect of a growing number of crystal defects, caused by the harsh oxidation conditions. Remarkably, the authors of [**13**] and [**14**] develop a "high pressure oxidation" method, suitable for the defect-free synthesis of overdoped cuprates. These works show that $T_c \approx 90$ K values can be obtained even in the p>0.6 region.

The results of [**13**] and [**14**] thus invalidate preceding high-temperature superconductivity theories by demonstrating that high-temperature superconductivity emerges from an ordinary Fermi-sea ground state. On the other hand, this result is well aligned with our model of electron Bose-Einstein condensation from ordinary Fermi-sea ground state.

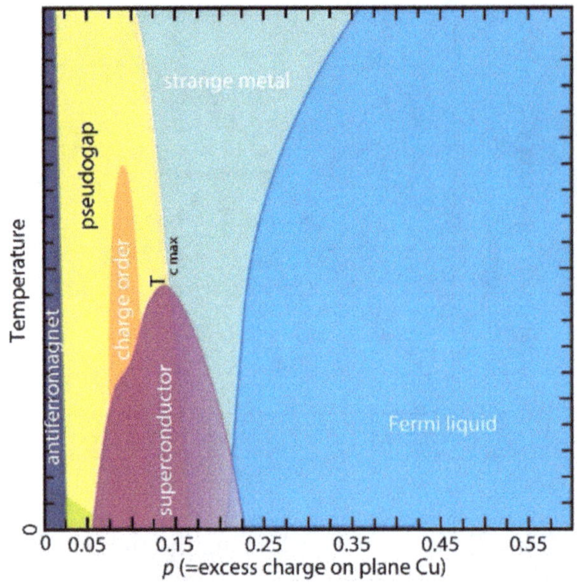

FIGURE 3.6.7. The historically formulated phase diagram of cuprate superconductors, which remained unchallenged till recent years. The superconducting phase is represented by the purple dome. The yellow area represents the so-called pseudogap phase, characterized by a reduced dimensionality of delocalized electron flows. The blue area represents the usual Fermi sea.

In section 3.5.3, we described the co-existence between Bose-Einstein condensed and Fermi-sea electron populations in the anisotropic case; such co-existence is indeed experimentally demonstrated in [**14**] for a highly overdoped superconductor. Our novel methodology is therefore suitable for the analysis of high-temperature superconductors such as $Ba_2CuO_{3.2}$, which are described as anisotropic rather than 2-dimensional [**13**].

3.6.5. Thin-film 2-dimensional superconductors. As explained in section 3.5.2, the T_{BEC} temperature of stacked 2-dimensional superconductors depends on the pre-condensation Fermi energy level. With all else being equal, a lower Fermi energy level implies proportionally lower T_{BEC} value.

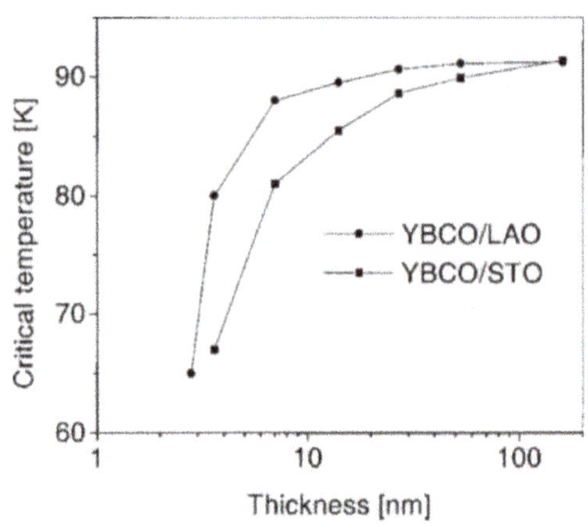

FIGURE 3.6.8. The T_c of $YBa_2Cu_3O_{7-\delta}$ nano-films, measured with two different substrate materials. Reproduced from [**15**].

As the thickness of a metal film reaches below the 100 nm range, the metal-substrate interface begins impacting the Fermi energy value. In other words, the Fermi energy level becomes dependent on the substrate material.

We consider the $YBa_2Cu_3O_{7-\delta}$ high-temperature superconductor material that comprises stacked 2-dimensional planes. As shown in figure 3.6.8, experimental measurements on a thin $YBa_2Cu_3O_{7-\delta}$ superconductor films indeed show that T_c becomes dependent on the substrate material as the film thickness reaches below the 100 nm range. This effect demonstrates the predicted T_c dependence on the pre-condensation Fermi energy level.

3.6.6. Fermi-level depletion in the superconducting state. Our model predicts a gradual Fermi energy level depletion in superconductors as their temperature is lowered towards zero; this is depletion is caused by the high electron occupancy of a Bose-Einstein condensate's energy eigenstates.

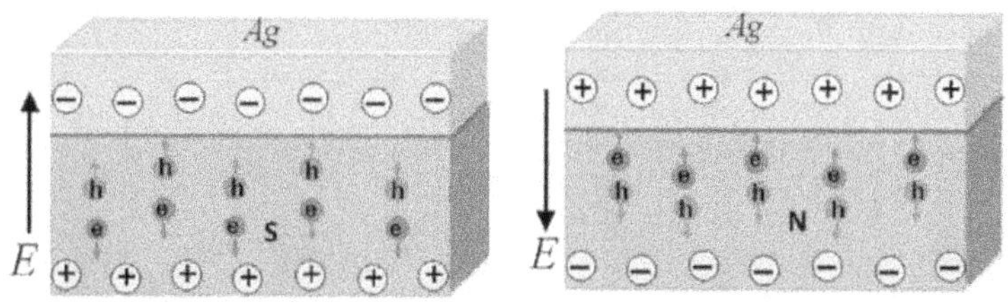

FIGURE 3.6.9. Fermi-level depletion upon transition from normal to superconducting state, manifesting as interface polarity change. Left: Ag-Superconductor interface in superconducting (S) state. Right: Ag-Superconductor interface in normal (N) state. The polarity is measured by laser-induced production of electron-hole pairs, which create photovoltaic effect.

A direct signature of this anticipated Fermi level change is reported in references [**16, 17, 18**]: these works measure MgB_2, $Bi-2223$, and $YBa_2Cu_3O_{7-\delta}$ materials' Fermi level relative to a silver contact. As shown in figure 3.6.9, the Fermi level depletion upon entering superconducting state manifests as a polarity change in the superconductor-silver interface. The interface polarity was detected by laser illumination technique, and the same effect was found for all superconducting materials.

In contrast, preceding superconductivity theories consider superconducting electrons to remain in the Fermi sea, i.e. they predict no Fermi level change upon superconducting state transition, and thus become contradicted by experiments.

3.7. Conclusions

We calculated the Bose-Einstein condensation temperature formulas for isotropic, stacked 2-dimensional, and anisotropic topologies. A remarkable match was found between the calculated temperature formulas and experimental data. We identified the coexistence of Fermi-Dirac and Bose-Einstein condensed particle populations in the isotropic and anisotropic cases, while the stacked 2-dimensional topology leads to a sharp phase transition.

We demonstrated that the superconducting phase transition is the Bose-Einstein condensation of the spin-triplet fraction of delocalized electrons. Our theory applies to both conventional and high-temperature superconductors.

We anticipate that the Bose-Einstein condensation temperature formulas derived in this work may lead to a generally accepted superconductivity theory based on conduction band electrons' Bose-Einstein condensation, and facilitate a rational search for higher-temperature superconductors.

Acknowledgements: The author thanks Marc Fleury (LENR Capital) for financially supporting a part of this research.

Bibliography

[1] The Nobel Prize in Physics 2001. https://www.nobelprize.org/prizes/physics/2001/popular-information/

[2] A. A. Shanenko et al "Mesoscopic samples: the superconducting condensate via the Gross–Pitaevskii scenario", Solid State Communications, volume 131.6, p. 409–414 (2004)

[3] F. London "On the Bose–Einstein Condensation", Physical Review, volume 54 .11, p. 947–952 (1938)

[4] C. J. Pethick and H. Smith "Bose–Einstein Condensation in Dilute Gases", Cambridge University Press (2001)

[5] H. M. Syed et al "Superconductivity in palladium hydride and deuteride at 52-61 kelvin", arXiv:1608.01774 (2016)

[6] Y. J. Uemura et al "Condensation, excitation, pairing, and superfluid density in high-T_c superconductors", Journal of Physics: Condensed Matter, volume 16.40 (2004)

[7] T. Klein et al "Metal-insulator transition and superconductivity in boron-doped diamond", Physical Review B, volume 75.16 (2007)

[8] H. P. Roeser et al "Electronic Energy Levels in High-Temperature Superconductors", Journal of Superconductivity and Novel Magnetism, volume 24, p. 1443-1451 (2011)

[9] H. Keller "Positive muons as probes in high-T_c superconductors", in "Exotic Atoms in Condensed Matter", Springer (1992)

[10] J.-L. Luo "The electronic state phase diagram of copper oxide high-temperature superconductors", p. 1-26 in "Advances in theoretical and experimental research of high temperature cuprate superconductivity", World Scientific (2020)

[11] J. Kačmarčik et al "Unusual interplay between superconductivity and field-induced charge order in $YBa_2Cu_3O_y$", Physical Review Letters, volume 121.16 (2018)

[12] Y. Itoh et al "^{31}P NMR studies of an iron-based superconductor $Ba_{0.5}Sr_{0.5}Fe_2(As_{1-x}P_x)_2$ with T_c=29 K", Journal of Physics: Conference Series, volume 1590 (2020)

[13] L. Sederholm et al "Extremely overdoped superconducting cuprates via high pressure oxygenation methods", Condensed Matter, volume 6 (2021)

[14] A. Gauzzi et al "Bulk superconductivity at 84 K in the strongly overdoped regime of cuprates", Physical Review B, volume 94.18 (2016)

[15] H. Y. Zhai et al "Effect of interfacial strain on critical temperature of $YBaCu_3O_{7-\delta}$ thin films", Applied Physics Letters, volume 76.23 (2000)

[16] Z. Chu et al "Photovoltaic effect in Ag/MgB_2 heterostructure", Journal of Alloys and Compounds, Volume 793 (2019)

[17] Z. Chu et al "Origin of photovoltaic effect in $(Bi, Pb)_2Sr_2Ca_2Cu_3O_{10+d}$/Ag heterostructure", Journal of Alloys and Compounds, volume 724 (2017)

[18] F. Yang et al "Polarity switching of the photo-induced voltage in YBCO/Ag heterojunction", Journal of Physics and Chemistry of Solids, volume 138 (2020)

[19] M. D. Watson et al "Multiband One-Dimensional Electronic Structure and Spectroscopic Signature of Tomonaga-Luttinger Liquid Behavior in $K_2Cr_3As_3$", Physical Review Letters, volume 118.9 (2017)

[20] S. Ogawa et al "Single Crystal Growth of and Hyperfine Couplings in the Spin-Triplet Superconductor $K_2Cr_3As_3$", Journal of the Physical Society of Japan, volume 92.6 (2023)

CHAPTER 4

Josephson frequency calculation

Andras Kovacs[1]
[1] ExaFuse
E-mail: andras.kovacs@broadbit.com

4.1. An overview of the Josephson effect

The Josephson effect is the oscillatory flow of electrons between two superconductors, that may occur when two superconductors are placed in proximity, with a thin insulating barrier between them. The Josephson effect is of practical interest because it exhibits a precise relationship between the involved voltage and frequency, both of which can be accurately measured.

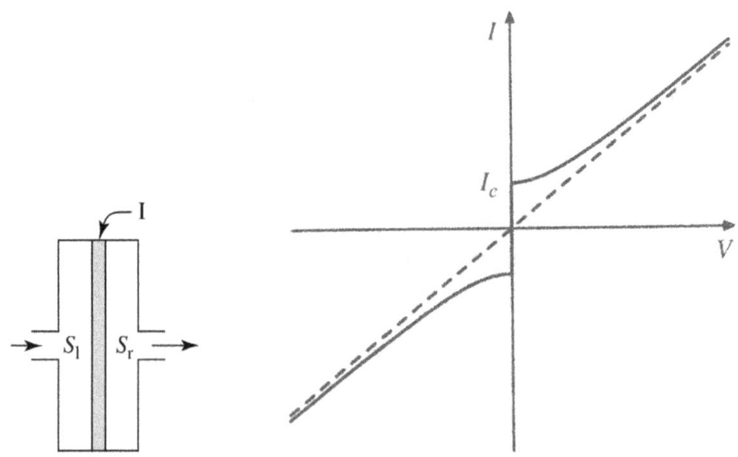

FIGURE 4.1.1. The superconducting electron flow across a Josephson junction is illustrated on the left side. The superconductors on the left (S_l) and right (S_r) carry the illustrated current flow, and are separated by a thin insulating film (I). The Josephson junction's $I-V$ characteristics is shown on the right side. An AC tunneling current is observed when $I > I_c$.

The arrangement of the two interfacing condensates is schematically shown in figure 4.1.1. The N electrons occupying a Bose-Einstein condensed ground state comprise a single wavefunction: $\psi = \sqrt{n}e^{i\varphi}$, where φ is the quantum mechanical phase of the Bose-Einstein condensed electron ground state and $n \equiv \frac{N}{V}$ is its electron density.

When the current flow is smaller than an I_c threshold value, there is zero electrostatic potential difference between the two superconductors. The current has a steady DC characteristic in this case; it is illustrated in the top part of figure 4.1.2. Since there is no potential difference across the insulator, let us consider the origin of electron tunneling asymmetry, that is required for the DC current flow. The tunneling probability is proportional to the ground state electron density, which we denote as n_l and n_r for the left and right sides, and therefore the $n_l > n_r$ condition results in the observed net DC current. One may ask how could the electron density vary, without causing an electrostatic potential difference. The Bose-Einstein condensate contains delocalized electrons at multiple

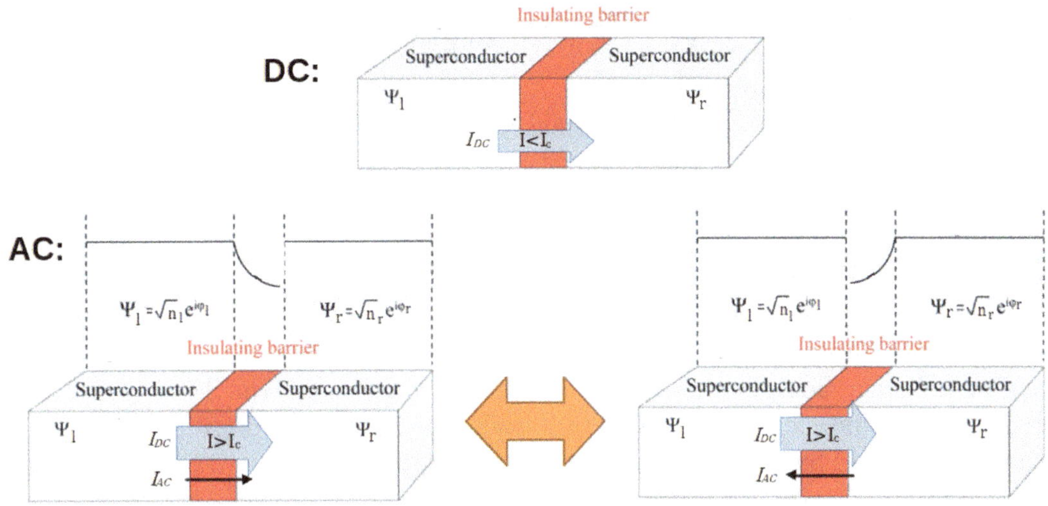

FIGURE 4.1.2. Top: both sides of the junction are at the same voltage when $I < I_c$, and the DC current flows steadily. Bottom: a voltage V appears across the junction when $I > I_c$, accompanied by an oscillating AC current component. The oscillation between the AC Josephson current's opposite phases is illustrated; the black arrow represents the AC current component. On the left side, the AC tunneling current originates fully from ψ_l. On the right side, the AC tunneling current originates fully from ψ_r. In-between, both ψ_l and ψ_r contribute to the tunneling current.

low-lying energy levels. In the DC Josephson effect context, raising the ground state electron density without impacting the electrostatic potential implies depleting the density of higher energy electron states, i.e. an adjustment of the Bose-Einstein condensed electron distribution. Such electron distribution adjustment is possible only up to a point, and that is the origin of the I_c current threshold.

When $I > I_c$, the difference between n_l and n_r is no longer compensated by other electrons, and a voltage drop V appears across the insulator. This is accompanied by the appearance of an oscillating tunneling current, generally referred to as the "AC Josephson current". This oscillating current is illustrated in the bottom part of figure 4.1.2, and we calculate its dynamics in the following paragraphs.

4.2. The calculation of Josephson oscillation frequency

Experiments show that the junction voltage V remains constant in the $I > I_c$ regime. Therefore, there is no displacement current contribution to the AC Josephson effect; it based on tunneling currents only.

Let us firstly consider the left side superconductor. We use the time-dependent Schrödinger equation to calculate the phase evolution of ψ_l. A Bose-Einstein condensed ground state has macroscopically large wavelength, and therefore its momentum goes to zero: $\hbar\frac{\partial\psi}{\partial x} \to 0$. The phase evolution of ψ is therefore determined by only two factors:

$$i\hbar\frac{\partial\psi_l}{\partial t} = eV\psi_l + K\psi_r \tag{4.2.1}$$

where ψ_l is the ground state wavefunction on the left side, ψ_r is the ground state wavefunction on the right side, and V is the electrostatic potential difference between the two sides.

As the insulator is thin, and the ground state accommodates a variable electron number, electrons tunnel between the ψ_l and ψ_r. The $K\psi_r$ term is the tunneling electrons' contribution to the phase evolution, and K is thus an unknown parameter. While we may write equation 4.2.1 with any reference potential level of our choice, this reference potential choice will influence the value of K. For equation 4.2.1, we choose the zero potential level to be the potential experienced by ψ_r, which determines the value of K. Consequently, the ψ_l is at eV potential in the context equation 4.2.1, and therefore it contains the $eV\psi_l$ term. We emphasize that the left side of equation 4.2.1 remains invariant under any reference potential choice; the factors in front of ψ_l and ψ_r vary with the reference potential choice, but the overall equation remains invariant.

We now consider the right side superconductor, and again use the Schrödinger equation to calculate the phase evolution of ψ_r. We choose the zero potential level to be the potential experienced by ψ_l: this is advantageous because by symmetry the cross-coupling term becomes $K\psi_l$, carrying the same K value as before. Consequently, the ψ_r is at $-eV$ potential in the context of our second Schrödinger equation, which takes the following form:

$$i\hbar\frac{\partial\psi_r}{\partial t} = -eV\psi_r + K\psi_l \tag{4.2.2}$$

We substitute $\psi_l = \sqrt{n_l}e^{i\varphi_l}$ and $\psi_r = \sqrt{n_r}e^{i\varphi_r}$ into equations 4.2.1 and 4.2.2:

$$i\hbar\left(\frac{\partial\sqrt{n_l}}{\partial t}e^{i\varphi_l} + i\frac{\partial\varphi_l}{\partial t}\sqrt{n_l}e^{i\varphi_l}\right) = eV\sqrt{n_l}e^{i\varphi_l} + K\sqrt{n_r}e^{i\varphi_r} \tag{4.2.3}$$

$$i\hbar\left(\frac{\partial\sqrt{n_r}}{\partial t}e^{i\varphi_r} + i\frac{\partial\varphi_r}{\partial t}\sqrt{n_r}e^{i\varphi_r}\right) = -eV\sqrt{n_r}e^{i\varphi_r} + K\sqrt{n_l}e^{i\varphi_l} \tag{4.2.4}$$

We rewrite the above equations, using $\frac{\partial\sqrt{n}}{\partial t} = \frac{1}{2\sqrt{n}}\frac{\partial n}{\partial t}$, then dividing by $\hbar e^{i\varphi_l}$ and $\hbar e^{i\varphi_r}$ respectively:

$$\frac{i}{2\sqrt{n_l}}\frac{\partial n_l}{\partial t} - \frac{\partial\varphi_l}{\partial t}\sqrt{n_l} = \frac{eV}{\hbar}\sqrt{n_l} + \frac{K}{\hbar}\sqrt{n_r}e^{i\phi} \tag{4.2.5}$$

$$\frac{i}{2\sqrt{n_r}}\frac{\partial n_r}{\partial t} - \frac{\partial\varphi_r}{\partial t}\sqrt{n_r} = -\frac{eV}{\hbar}\sqrt{n_r} + \frac{K}{\hbar}\sqrt{n_l}e^{-i\phi} \tag{4.2.6}$$

where $\phi \equiv \varphi_r - \varphi_l$.

In the above equations, both the real and imaginary parts must match each other. The imaginary parts give $\frac{\partial n_l}{\partial t} = \frac{2K}{\hbar}\sqrt{n_l n_r}\sin\phi$ and $\frac{\partial n_r}{\partial t} = -\frac{2K}{\hbar}\sqrt{n_l n_r}\sin\phi$. Therefore, $\frac{\partial n_l}{\partial t} = -\frac{\partial n_r}{\partial t}$ which makes sense because the number of electrons gained on the left side are lost from the right side, and vice versa. As $-e\frac{\partial n}{\partial t}$ gives the current density across the Josephson junction, it follows that a sinusoidally oscillating current appears:

$$J = e\frac{2K}{\hbar}\sqrt{n_l n_r}\sin\phi \tag{4.2.7}$$

The real parts of equations 4.2.5-4.2.6 give $-\frac{\partial\varphi_l}{\partial t} = \frac{eV}{\hbar} + \frac{K}{\hbar}\sqrt{\frac{n_r}{n_l}}\cos\phi$ and $-\frac{\partial\varphi_r}{\partial t} = -\frac{eV}{\hbar} + \frac{K}{\hbar}\sqrt{\frac{n_l}{n_r}}\cos\phi$. With the same superconducting material on both sides of the junction, we anticipate $n_l \approx n_r$. Although the DC current flow requires $n_l > n_r$, their difference is a tiny fraction, and thus $\frac{n_r}{n_l} \to 1$. Subtracting the real parts of equations 4.2.5-4.2.6 then yields the Josephson frequency of the junction current:

$$\omega_{Josephson} = \frac{\partial \phi}{\partial t} = \frac{\partial \varphi_r}{\partial t} - \frac{\partial \varphi_l}{\partial t} = \frac{2e}{\hbar} V \tag{4.2.8}$$

One may ask whether it is allowed to subtract equations 4.2.5-4.2.6, which were formulated using different reference potential values. The key point is to recognize the physical observables, that do not depend on the choice of reference potential. For instance, n_l and n_r are measurable physical observables, that are independent of our reference potential choice. The left sides of equations 4.2.5 and 4.2.6 define the time evolution of the system: for each equation its time evolution is a physical observable because the emitted radiation frequency is a physical observable. In other words, if $\frac{\partial \varphi_l}{\partial t}$ depended on our choice of reference potential, the emitted radiation frequency would be no longer well defined. Since the left sides of equations 4.2.5 and 4.2.6 are independent of our reference potential choice, their right sides must be also independent of our reference potential choice. While the individual terms on the right sides of equations 4.2.5 and 4.2.6 explicitly depend on our reference potential choice, their sum is invariant. Therefore, our goal was to find such a practical reference potential choice for each equation that makes our overall calculation as simple as possible. In summary, the subtraction of equations 4.2.5-4.2.6 is a perfectly valid operation.

A V=1 μV potential difference across the junction leads to $f = \frac{\omega_{Josephson}}{2\pi}$=483.6 MHz current oscillation frequency, which must emit radiation at the same frequency. This frequency range is measurable by radio frequency equipment.

4.3. Errors in preceding Josephson frequency calculations

We point out three crucial paradoxes in preceding superconductivity theories' interpretation of the Josephson effect. Firstly, preceding theories consider superconducting electrons to be at the Fermi energy level, while at the same time paradoxically using the zero-momentum equations 4.2.1 and 4.2.2. Secondly, preceding theories consider numerous superconducting electron pairs condensing into the same wavefunction, as required for equations 4.2.1 and 4.2.2, while at the same time paradoxically claiming that this is not a Bose-Einstein condensate. Thirdly, preceding theories consider equations 4.2.1 and 4.2.2 with a common reference potential, set to the potential value at the mid-point of the insulating film. This means that ψ_l and ψ_r experience different electric potentials, and therefore the cross-coupling term of equations 4.2.1 and 4.2.2 cannot carry the same K value. In summary, the Josephson effect naturally matches electrons' Bose-Einstein condensation model, but becomes paradoxical in the context of preceding superconductivity theories.

4.4. Andreev reflections

We address the phenomenon of so-called Andreev reflections, which involve electron transition from a semiconducting material into a superconducting metal. During an Andreev reflection, a single electron travels towards a semiconductor-superconductor interface; it transfers $2e$ charge across the interface, leaving behind a positively charged hole that travels in opposite direction with respect to the incoming electron. The transferred $2e$ charges have anti-parallel correlated spin.

Some superconductivity theorists claim that Andreev reflections arise because superconducting electrons are in a paired state, and that electron transition into a superconducting phase requires $2e$ charge transfer. But such claims stand in contrast to the above calculated results. Given that the Fermi-Dirac and Bose-Einstein condensed electron phases co-exist in a superconductor, let us firstly clarify whether the $2e$ charge is transferred into the Fermi-Dirac or the Bose-Einstein phase. In Andreev reflection experiments, the semiconductor's Fermi level is at lower energy than the superconductor's Fermi level at $T > T_c$, so that the incoming electron pair has too low energy to occupy an available electron state in the superconducting material at $T > T_c$ - i.e. when it is in its normal state. With the help of such set-up, some superconductivity researchers aim to obtain signatures of "weakly bound but traveling" electron pairs, assuming that a hypothetical "Cooper-pair binding energy" moves the electron pair across the apparent energy barrier. The hidden assumption here is that Fermi energy level remains invariant between the normal and superconducting states; however the experimental data of section 3.6.6 shows that this assumption is absolutely false. The Fermi energy level is depleted as the temperature is lowered below T_c, and this removes the energy barrier against electron entry into the superconductor's Fermi-sea phase.

Since the Andreev reflection involves hole creation in the semiconducting material, the injected traveling electron forms and incoherent singlet pair with an electron of the semiconductor material. The spin-singlet state is inferred from their anti-parallel spin correlation. Such a spin-singlet electron pair can freely transition to the superconductor's Fermi-sea phase, given that there is no energy barrier between the two Fermi energy levels. On the other hand, a spin-singlet electron pair cannot directly transition into the Bose-Einstein condensed electron phase, for the reasons explained in chapter 3. Therefore, Andreev reflections do not involve any interaction with the Bose-Einstein condensed electron phase; they are unrelated to superconductivity phenomena.

4.5. Experimental validations

The $\frac{2e}{\hbar}$ factor of equation 4.2.8 is also called the Josephson constant K_J. Using voltage and radiation frequency measurements, equation 4.2.8 has been validated at a great precision; the current CODATA value for K_J is $4.835978484 \cdot 10^{14}$ Hz/V [**1**].

The Schrödinger equations of section 4.1 are not specific to electrons, and therefore the same effect should arise also for other types of Bose-Einstein condensates. As anticipated, references [**2, 3**] report Josephson effect observations for atomic Bose-Einstein condensates.

In section 3.6.2, we showed experimental evidence for individual electrons' transition into the superconducting Bose-Einstein condensate. This data validates our methodology of calculating the transition of individual electrons between the two Bose-Einstein condensates.

In summary, our Josephson frequency calculation proves yet again that superconducting electrons form a Bose-Einstein condensate.

Bibliography

[1] "2022 CODATA Value: Josephson constant", The NIST Reference on Constants, Units, and Uncertainty (2022)

[2] S. Levy et al "The ac and dc Josephson effects in a Bose–Einstein condensate", Nature, volume 449.7162 (2007)

[3] R. Gati et al "Realization of a single Josephson junction for Bose–Einstein condensates", Applied Physics B, volume 82 (2006)

CHAPTER 5

Catalysts of electrons' Bose-Einstein condensation

Andras Kovacs[1]
[1] ExaFuse
E-mail: andras.kovacs@broadbit.com

5.1. Simple examples of catalyzed Bose-Einstein condensation

Having explored the thermodynamics of Bose-Einstein condensation, the remaining challenge is to predict the peak T_c temperature for various materials. To make progress towards this practical goal, we must identify the catalytic effects on Bose-Einstein condensation, which increase the fraction of spin-triplet electron pairs. A catalyst can be embedded in the material itself, or can be externally applied. For any given material, the T_c prediction would firstly estimate the energy difference between spin-singlet and spin-triplet states of conduction band electron pairs, then estimate the effect of an applied catalyst on the ratio of spin-triplet electrons, and finally calculate T_c by applying chapter 3 formulas. To illustrate the effect of various catalysts, we start from simple examples.

The photo-illumination experiments described in reference **[1]** present a striking example of UV illumination's catalytic effect on T_c. The authors of **[1]** illuminated an under-doped $YBa_2Cu_3O_{7-\delta}$ superconductor for some hours, and half an hour after the illumination measured up to 13% higher T_c than before the experiment. The T_c increase was higher for 300 nm UV illumination than for visible light illumination. Light illumination can therefore increase the fraction of Bose-Einstein condensation capable electrons.

5.2. Phonons' catalytic role in elementary metals

At this point, we can identify the true role of phonons in superconductivity. Table 1 lists the highest T_c elementary metals at ambient pressure. Elementary metals in the T_c>2 K group are either single-isotope elements or display the $T_c \sim M^{-0.5}$ isotope effect, where M is the isotope mass. It is therefore probable that all metals of the T_c>2 K group would display the $T_c \sim M^{-0.5}$ isotope effect, if various isotopes of each element were available. The $T_c \sim M^{-0.5}$ isotope effect is associated with phonon interactions for two reasons: i) the Debye frequency of phonons is also proportional to $M^{-0.5}$, and ii) except for electron-phonon interactions, the nuclear mass plays no role in delocalized electron dynamics.

In contrast, for elementary metals in the T_c<2 K group, the isotope mass has a randomly varying coefficient and there is no isotope effect at all in metals such as Ru or Zr. I.e. phonons play no essential role in elementary metals' superconductivity in the T_c<2 K group.

With reference to Homes' law, phonons have a mildly catalytic effect by scattering delocalized electrons. In a small fraction of electron pairs, the phonon-induced scattering flips their spin correlation into spin-triplet state. Without this phonon effect, elementary metals cannot reach even 2 K superconductivity at ambient pressure.

The phenomenology of catalytic phonon effects is described in reference **[2]**, quantifying the impact of electron-phonon coupling. The experimental data reviewed in reference

Metal	T_c	Isotope effect
Nb	9.25 K	
Tc	8.2 K	
Pb	7.2 K	$T_c \sim M^{-0.48}$
La	6 K	
V	5.4 K	
Ta	4.4 K	
Hg	4.15 K	$T_c \sim M^{-0.5}$
Sn	3.7 K	$T_c \sim M^{-0.46}$
In	3.4 K	
Tl	2.4 K	$T_c \sim M^{-0.5}$
Re	1.7 K	$T_c \sim M^{-0.38}$

TABLE 1. The listing of highest T_c elementary metals at ambient pressure. The last column shows the isotope effect measurement, if the given element has various isotopes.

[2] shows that a T_c of around 20 K can be achieved via phonon-catalyzed superconductivity.

5.3. The omnipresent coincidence between peak T_c and a phase change

A systematic analysis of superconducting phase diagrams reveals that superconducting materials' phase changes have a catalytic effect on their electrons' Bose-Einstein condensation. This is an independent catalytic effect from the above discussed phonon effect.

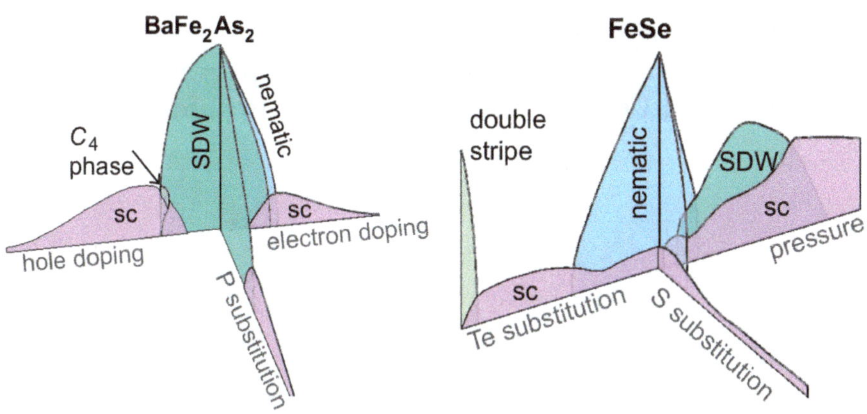

FIGURE 5.3.1. The approximate coincidence between a phase-change temperature and the maximum temperature of a superconducting dome in iron-based $BaFe_2As_2$ and FeSe type superconductors. Reproduced from [3].

Figures 5.3.1-5.3.5 illustrate the catalytic effect of phase changes: the T_c peak location approximately coincides with a phase-change location. This effect is observed across many different superconductor families. As shown in figure 5.3.1, variations of this same effect can be observed even within one specific superconductor type, by varying the doping type. This universally occurring coincidence between the T_c peak location and phase-change location was only discovered in the 21st century, and was not predicted by any preceding superconductivity theory.

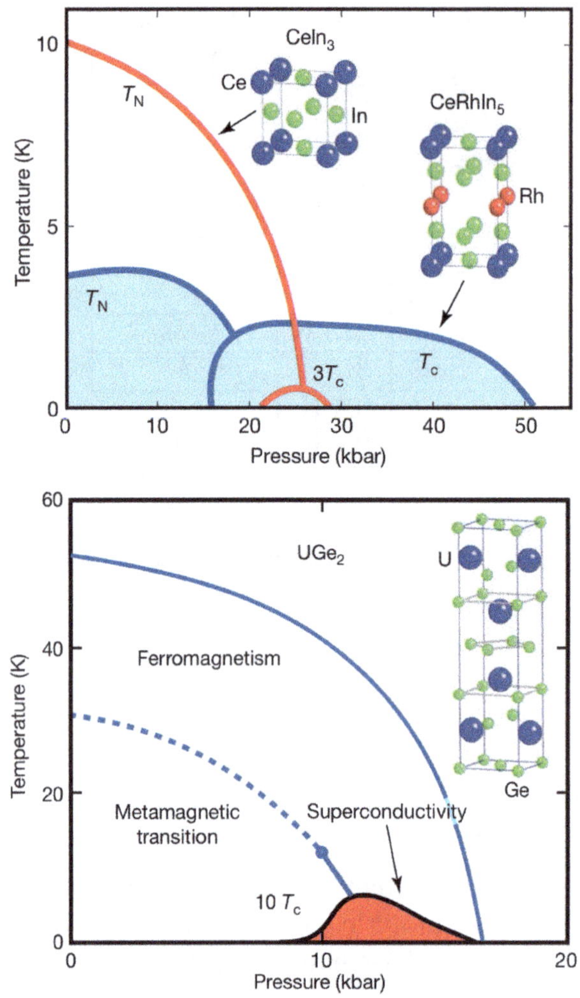

FIGURE 5.3.2. The approximate coincidence between a phase-change temperature and the maximum temperature of a superconducting dome in $CeIn_3$, $CeRhIn_5$ and UGe_2 type superconductors. Reproduced from [**4**].

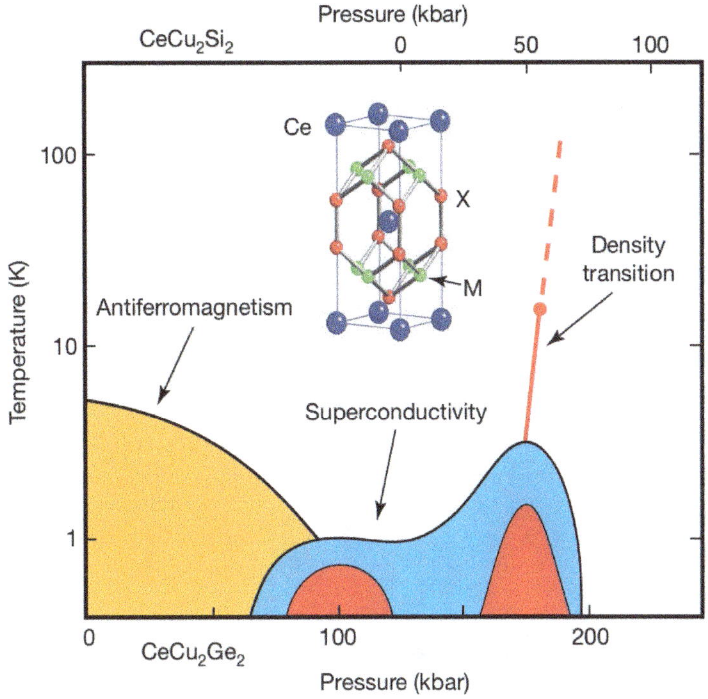

FIGURE 5.3.3. The approximate coincidence between a phase-change temperature and the maximum temperature of a superconducting dome in $CeCu_2Si_2$ and $CeCu_2Ge_2$ type superconductors. Reproduced from [**4**].

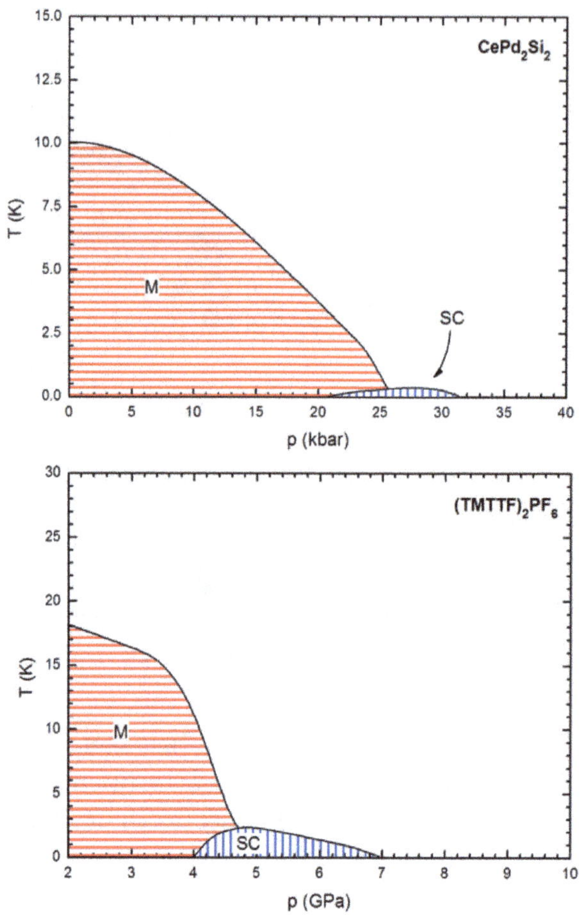

FIGURE 5.3.4. The approximate coincidence between a phase-change temperature and the maximum temperature of a superconducting dome in $CePd_2Si_2$ superconductor and in an organic salt superconductor. Reproduced from [**5**].

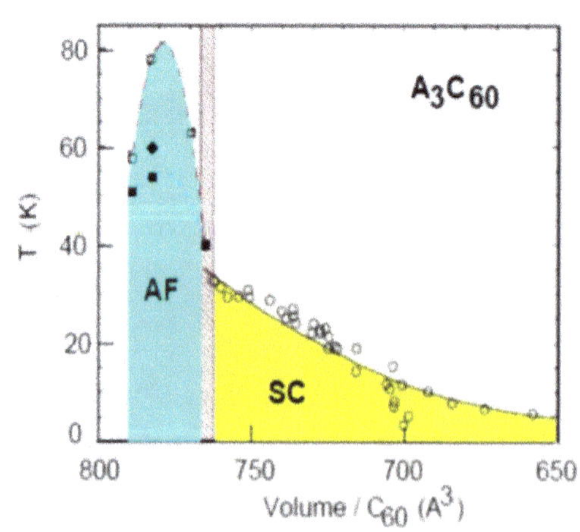

FIGURE 5.3.5. The approximate coincidence between a phase-change temperature and the maximum temperature of a superconductivity in Buckminsterfullerene-based superconductors. Reproduced from [**6**].

A phase change involves the re-arrangement of electron orbitals or lattice structure. Close to the phase change temperature, there is some co-existence of two phases. Such co-existence involves a dynamic re-arrangement of lattice structures: these dynamic re-arrangements create electron scattering sites that increase the resistivity of the normal state. With reference to Homes' law, being near the phase-change boundary catalyzes electrons' Bose-Einstein condensation. Specifically, the phase-change induced disorder causes scattering in a fraction of electron pairs; this process flips their spin correlation into spin-triplet state.

Regarding cuprate type superconductors, while most of them have a single superconducting dome in the hole-doping region, there are examples of double superconducting domes. $La_{2-x}Ba_xCuO_4$ is the most well-known such superconductor. After its discovery, many speculative theories were published about mechanisms which "suppress superconductivity around the x=0.12 doping region". However, our results reveal that nothing suppresses superconductivity around the x=0.12 doping region: the two distinct regions of optimal catalytic effects are around x=0.08 and x=0.16, and the Bose-Einstein condensation catalyst is not strong enough to overlap the two domes. If the $La_{2-x}Ba_xCuO_4$ material had stronger catalysts of Bose-Einstein condensation, leading to higher T_c, the two domes would nicely overlap and there would be superconductivity also around the x=0.12 doping region. This process is nicely illustrated by the $La_{2-x}Sr_xCuO_4$ and $YBa_2Cu_3O_{7-\delta}$ superconductors; under a sufficiently strong magnetic field, which works against the catalysis of electrons' Bose-Einstein condensation, the $La_{2-x}Sr_xCuO_4$ and $YBa_2Cu_3O_{7-\delta}$ phase diagrams also have two distinct superconducting domes [**7, 8**]. Without an applied magnetic field, the two superconducting domes of $La_{2-x}Sr_xCuO_4$ and $YBa_2Cu_3O_{7-\delta}$ overlap, and their phase diagram is dominated by the larger superconducting dome.

All these observations lead up to a major question: what is the phase-change at the peaks of cuprate materials' superconducting domes, which catalyzes their exceptionally high T_c? Taking the $YBa_2Cu_3O_{7-\delta}$ example, which is a representative hole-doped cuprate superconductor, figure 5.3.6 shows that both of its superconducting dome peaks coincide with the onset of a "charge density wave" (CDW) phase. Figure 5.3.7 shows the same coincidence for the $La_{2-x}Ba_xCuO_4$ superconductor. The two-domed phase diagram of hole-doped cuprates such as $La_{2-x}Ba_xCuO_4$, $La_{2-x}Sr_xCuO_4$, or $YBa_2Cu_3O_{7-\delta}$ is naturally explained by the two legs of the CDW phase-change, which define two catalytic regimes.

The CDW phase is a periodic lattice distortion, which occurs spontaneously below T_{CDW}. In materials that comprise a CDW phase, the presence of such lattice distortion lowers the total energy of delocalized electrons. While superconductivity researchers notice that the CDW phase occurs in several cuprate type superconductors, this phase transition's catalytic role has not been understood in prior studies. Up to now, figure 5.3.7 was interpreted to mean that the CDW and superconductivity mechanisms are "in competition" [**7**], since T_{CDW} reaches its maximum in the same region where T_c reaches its minimum. In contrast, we recognize a catalytic effect all along the phase-change boundary denoted by the T_{CDW} curve, and it is chapter 3 equations that determine how high the superconducting domes are pulled up along the two legs of the T_{CDW} curve.

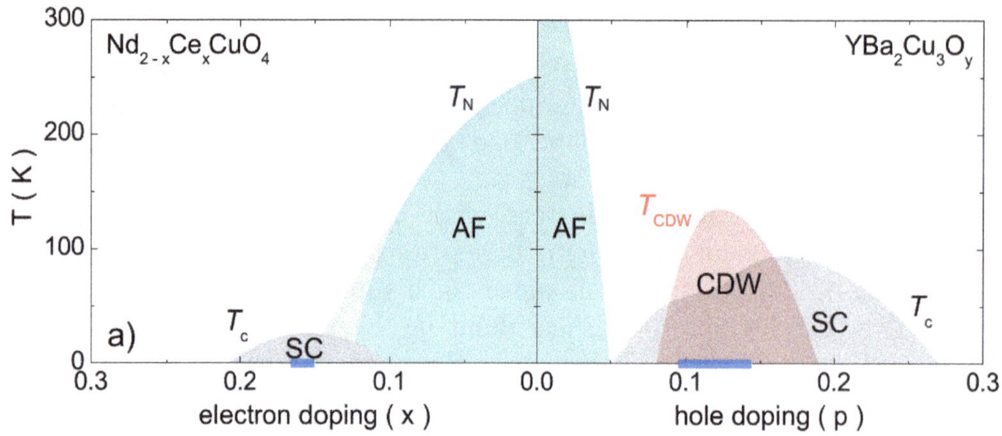

FIGURE 5.3.6. The phase diagram of electron-doped and hole-doped cuprate superconductors. The CDW phase-change temperature and the maximum temperature of superconducting dome peaks coincide in $YBa_2Cu_3O_{7-\delta}$.

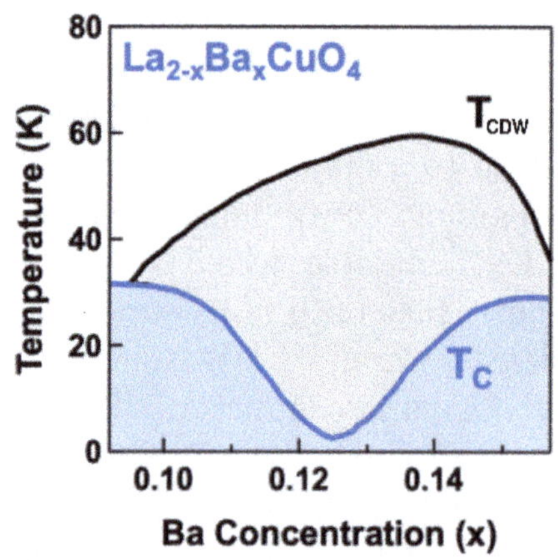

FIGURE 5.3.7. The CDW phase-change temperature (T_{CDW}) and the maximum temperature of superconducting dome peaks ($T_{c,max}$) coincide also in $La_{2-x}Ba_xCuO_4$.

Having explored the catalytic role of CDW phase-changes, we are left with the following question: does the CDW phase-change related catalytic effect explain hole-doped cuprates' exceptionally high T_c? Or is some other catalytic mechanism responsible for hole-doped cuprates' exceptionally high T_c? In the following chapter, we explore this question.

Bibliography

[1] R. El Hage et al "Disentangling photodoping, photoconductivity, and photosuperconductivity in the cuprates", Physical Review Letters, Volume 132.6 (2024)

[2] M. Surma "A New Semiempirical Formula for Superconducting Transition Temperatures of Metals and Alloys", Phys. Stat Sol., Volume 116 (1983)

[3] R. M. Fernandez et al "Ironing out the details of unconventional superconductivity", arXiv:2201.02095 (2022)

[4] P. Monthoux et al "Superconductivity without phonons", Nature Reviews, Volume 450 (2007)

[5] A. Martinelli et al "The phase diagrams of iron-based superconductors: Theory and experiments", Comptes Rendus Physique, Volume 17 (2016)

[6] Y. J. Uemura "Energy-scale considerations of unconventional superconductors—implications to condensation and pairing", Physica C, Volume 614 (2023)

[7] S. Caprara et al "Doping-dependent competition between superconductivity and polycrystalline charge density waves", SciPost Physics, Volume 8.1 (2020)

[8] B. J. Ramshaw et al "Quasiparticle mass enhancement approaching optimal doping in a high-T_c superconductor", Science, Volume 348 (2015)

CHAPTER 6

Weakly bound electrons in doped metals: a predicted electron state that catalyzes spin-triplet production

Andras Kovacs[1]
[1] ExaFuse
E-mail: andras.kovacs@broadbit.com

ABSTRACT. While the preceding chapters derived experimentally validated relationships, this chapter makes predictions that are based on already established results. We predict an electron state that allows spin coherent electrons at relatively high temperatures. Specifically, we predict the existence of weakly bound electron orbitals in hole-doped metals, and calculate their binding energy. The binding energy of such electron orbitals is in the 10 meV range, and their mean radius is in the nm range. Their existence follows from relatively simple calculations.

Regarding hole-doped cuprate materials, our calculations show that the limiting temperature of a weakly bound electron state is similar to hole-doped cuprates' peak T_c. It is possible that this newly recognized electron state catalyzes conduction band electrons' Bose-Einstein condensation.

6.1. Introduction

Recall from chapter 3 that electron pairs' transition between spin-singlet and spin-triplet states is a pre-condition to Bose-Einstein condensation: this pre-condition is illustrated in figure 3.1.1. Weaker spin pairing energies facilitate more probable transition into spin-triplet state. An electron pair's spin pairing energy is determined by their proximity. In the context of bound electrons, the larger their mean orbital radius is, the weaker their spin pairing energy becomes. It follows that the fraction of spin-triplet electron pairs is the highest in weakly bound electron states. In other words, a weakly bound electron state is the best candidate for observing bound electrons' spin coherence.

High-temperature superconductors are hole-doped metals. Figure 2.1.1 shows their residual diamagnetism extending up to twice their superconducting T_c temperature. The same phenomenon is seen also in other doped metals, such as $Nb_{0.15}Si_{0.85}$. In contrast, figure 2.1.1 shows ordinary lead metal's residual diamagnetism fading out very soon above its superconducting T_c temperature. This data suggests that hole-doped metals may host a coherent electron state, whose coherence is maintained for some extended half-life.

Altogether, these observations motivate the study of weakly bound electron states in hole doped metals.

6.2. Thomas-Fermi screening

We start by deriving the Thomas-Fermi screening mechanism, which is well-established and applicable to all doped metals. Suppose that one lattice ion of a metal is replaced by a more positively charged ion. How do delocalized electrons respond to this more positive site? As they are attracted to it, the electron density increases around this more positive site.

In the strict sense of the Drude-Sommerfeld model, which was introduced in section 3.2, single frequency electron waves uniformly fill the the interior of a metal, and thus cannot respond to any local positive charge. We need to employ a careful approximation, which allows to estimate the local electron response without giving up the Drude-Sommerfeld model.

The main idea is illustrated in figure 6.2.1. The zero level of figure 6.2.1 corresponds to the lowest k value in the interior of the metal, and the horizontal lines above it are the higher k value states. The potential dip around the positive charge shifts down this lowest energy state; the delocalized electrons are attracted to the more positively charged region. Since the electrons minimize their total energy, the top of the electron sea remains flat at the Fermi energy level.

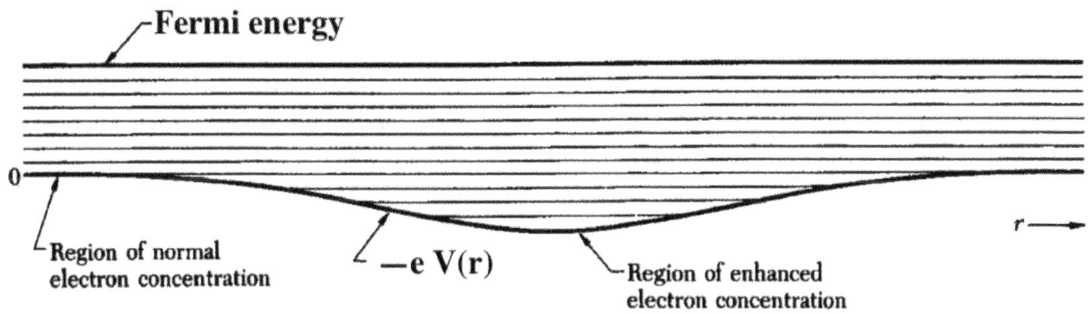

FIGURE 6.2.1. The electron screening around a positive charge is modeled as additional standing waves at a lower energy level around that charge

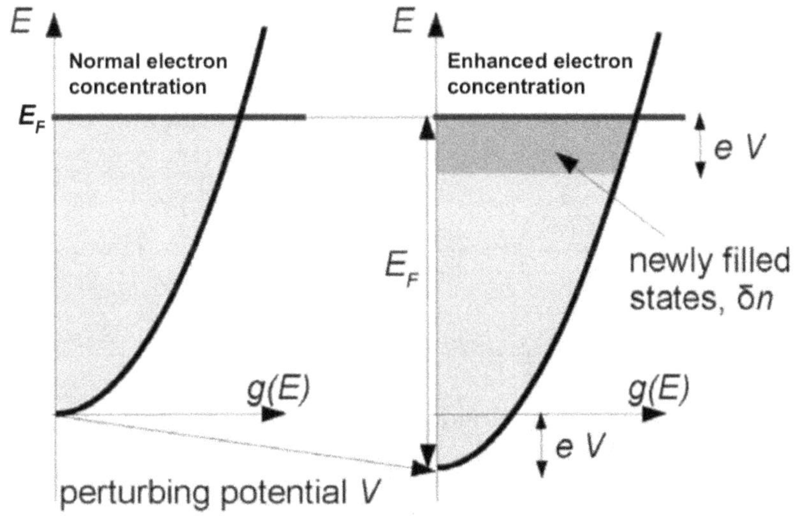

FIGURE 6.2.2. The appearance of newly filled electron states near the positive charge

Although delocalized electrons are single-frequency standing waves only in the region of normal electron concentration, the simplifying assumption of the Thomas-Fermi screening theory is to model them as single frequency standing waves also in the sloping region of figure 6.2.1. Then the net effect is the appearance of additional electron states, as shown in the pink shaded region of figure 6.2.2.

An other way of thinking about the Thomas-Fermi screening theory is to imagine that the potential slope of figure 6.2.1 is approximated by a staircase-like sequence of flat segments, which are in thermodynamic balance with each other. Notice that the Fermi energy level has higher k_F in the segments which are closer to the positive charge than in the segments of normal electron concentration. This reflects the dynamic that delocalized electrons speed up in the proximity of a positive charge.

Our objective is to calculate the $V(r)$ potential curve, which corresponds to the equilibrium distribution of delocalized electrons. It comprises two parts:

$$V(r) = V_{ext}(r) + \delta V(r)$$

where $V_{ext}(r) = \frac{1}{4\pi\varepsilon_0}\frac{Z_{ex}e}{r}$ is the unscreened Coulomb potential of the $Z_{ex}e$ positive excess charge, and the $\delta V(r)$ screening potential is calculated from the increased electron density by using Poisson's equation:

$$\nabla^2 \delta V(r) = \frac{e}{\varepsilon_0}\delta n = \frac{e}{\varepsilon_0} 2d_{3d}(E_F)\left[eV(r)\right] \tag{6.2.1}$$

where $d_{3d}(E_F)$ is the density of electron states given by equation 3.2.8, and the factor of 2 accounts for the pair-wise electron occupancy of each Fermi sea state. The $eV(r)$ energy difference is graphically illustrated in figure 6.2.2.

We guess that the solution is of the following form:

$$V(r) = \frac{1}{4\pi\varepsilon_0}\frac{Z_{ex}e}{r}e^{-rk_{TF}} \tag{6.2.2}$$

where k_{TF} is the Thomas-Fermi wavenumber, whose $r_{TF} = k_{TF}^{-1}$ inverse is also referred to as the Thomas-Fermi screening length. The resulting expression for $\delta V(r)$ is then the following:

$$\delta V(r) = \frac{1}{4\pi\varepsilon_0}\frac{Z_{ex}e}{r}\left(e^{-rk_{TF}} - 1\right) \tag{6.2.3}$$

In spherical coordinates, the left side of equation 6.2.1 evaluates to:

$$\left[\nabla^2 \delta V(r)\right]_{3d} = \frac{1}{r^2}\frac{\partial}{\partial r}\left(r^2 \frac{\partial \delta V(r)}{\partial r}\right) = \frac{Z_{ex}e}{4\pi\varepsilon_0}\frac{1}{r^2}\frac{\partial}{\partial r}\left(-\left(e^{-rk_{TF}} - 1\right) - rk_{TF}\left(e^{-rk_{TF}}\right)\right) =$$
$$= \frac{Z_{ex}e}{4\pi\varepsilon_0}\left[\frac{k_{TF}}{r^2}e^{-rk_{TF}} - \frac{k_{TF}}{r^2}e^{-rk_{TF}} + \frac{k_{TF}^2}{r}e^{-rk_{TF}}\right] = \frac{k_{TF}^2}{4\pi\varepsilon_0}\frac{Z_{ex}e}{r}e^{-rk_{TF}} \tag{6.2.4}$$

In the 3-dimensional case, the right side of equation 6.2.1 evaluates to:

$$\frac{e^2}{\varepsilon_0}2d_{3d}(E_F)\frac{1}{4\pi\varepsilon_0}\frac{Z_{ex}e}{r}e^{-rk_{TF}} = \left(\frac{e^2}{4\pi\varepsilon_0^2}\left(\frac{\sqrt{m}}{\hbar}\right)^3 \frac{\sqrt{2}}{\pi^2}\sqrt{E_F}\right)\frac{Z_{ex}e}{r}e^{-rk_{TF}} \tag{6.2.5}$$

The correspondence between the above two equations shows that we guessed the correct form of the solution, and the fulfillment of equation 6.2.1 yields the following result for k_{TF}:

$$\left[k_{TF} = e\sqrt{\left(\frac{\sqrt{m}}{\hbar}\right)^3 \frac{\sqrt{2}}{\varepsilon_0\pi^2}\sqrt[4]{E_F}}\right]_{3d} \tag{6.2.6}$$

In common metals, $E_F \approx$7 eV. Since k_{TF} only weakly depends on the precise E_F value, we can evaluate k_{TF} by using E_F=7 eV. The resulting Thomas-Fermi screening length is:

$$\left[k_{TF} \approx 18 \times 10^9\, m^{-1}\right]_{3d} \tag{6.2.7}$$

$$\left[r_{TF} \approx 0.056\, nm\right]_{3d}$$

In the 2-dimensional case, the left side of equation 6.2.1 is difficult to evaluate for a single plane, because the charge distribution is no longer spherically symmetric. In practice, 2-dimensional metals comprise stacked planes, with graphite-like structure. For stacked planes, we can estimate that the screening charge distribution is still spherically symmetric, with the strongest screening in the plane where the positively charged ion is, and gradually weaker screening in the neighboring planes. The left side of equation 6.2.1 then evaluates in the same way as for the 3-dimensional case:

$$\left[\nabla^2 \delta V(r)\right]_{2d,stack} \approx \left[\nabla^2 \delta V(r)\right]_{3d} \tag{6.2.8}$$

We proceed to evaluate the right side of equation 6.2.1:

$$\frac{e^2}{\varepsilon_0}\left[2d_{2d}(E_F)\frac{1}{d_{il}}\right]\frac{1}{4\pi\varepsilon_0}\frac{Z_{ex}e}{r}e^{-rk_{TF}} = \left(\frac{e^2}{4\pi\varepsilon_0^2}\frac{m}{\hbar^2}\frac{1}{\pi d_{il}}\right)\frac{Z_{ex}e}{r}e^{-rk_{TF}} \tag{6.2.9}$$

where d_{il} is the inter-layer spacing of the 2-dimensional stack, and $d_{2d}(E_F)$ is given by equation 3.2.9. As $2d_{2d}(E_F)$ counts the density of states within one plane, we need to use the $2d_{2d}(E_F)\frac{1}{d_{il}}$ expression in order to get the volumetric density of states.

The above equations lead to the following result for k_{TF}:

$$\left[k_{TF} = e\sqrt{\frac{m}{\varepsilon_0\hbar^2}\frac{1}{\pi d_{il}}}\right]_{2d,stack} \tag{6.2.10}$$

In these 2-dimensional stacked materials, the Thomas-Fermi screening length mainly depends on the inter-layer spacing parameter. In graphite, d_{il} =0.34 nm, and we thus get $k_{TF} = 2.4 \times 10^9\, m^{-1}$ and r_{TF}=0.4 nm. That is an order of magnitude larger screening radius value than in ordinary 3-dimensional metals. The most studied high-temperature superconductor material is $YBa_2Cu_3O_{7-\delta}$; due to its highly anisotropic diamagnetism it can be modeled as a stack of 2-dimensional metallic planes. According to measurements, $YBa_2Cu_3O_{7-\delta}$ has 1.17 nm lattice parameter along its c-axis, which contains two conducting planes. Therefore, d_{il} =0.585 nm in $YBa_2Cu_3O_{7-\delta}$, and its Thomas-Fermi screening length is r_{TF}=0.52 nm, which is slightly larger than in graphite.

6.3. Orbitals under electron screening effect

A natural question is to ask what prevents some delocalized electrons from forming a localized orbit around a more positively charged doping site. The main hindrance is that any unbound electron sees only a fraction of this larger nuclear charge, as it is screened by other delocalized electrons.

Suppose that an electron does form a localized Bohr orbit around a Z_{ex} charged site, at a mean orbit radius r_{scr}, where the *scr* index denotes that it is a partially screened orbit. We calculate the positive charge which the localized electron experiences at the r_{scr} distance. In the unscreened case, an electron localized at mean distance r experiences $\frac{1}{4\pi\varepsilon_0}\frac{Z_{ex}e}{r}$ potential and $\frac{-1}{4\pi\varepsilon_0}\frac{Z_{ex}e}{r^2}$ electric field. In the actual screened case, the localized electron experiences the following potential:

$$U = \frac{1}{4\pi\varepsilon_0}\frac{Z_{ex}e}{r_{scr}}e^{-r_{scr}/r_{TF}}$$

The experienced electric field is the radial derivative of the screened potential:

$$E = \frac{-1}{4\pi\varepsilon_0}\frac{Z_{ex}e}{r_{scr}}e^{-r_{scr}/r_{TF}}\left(\frac{1}{r_{scr}}+\frac{1}{r_{TF}}\right)$$

The ratio of screened and unscreened charges corresponds to the ratio of screened and unscreened electric fields:

$$Z_{scr} = Z_{ex}e^{-r_{scr}/r_{TF}}\left(1+\frac{r_{scr}}{r_{TF}}\right) \tag{6.3.1}$$

At the same time, in the eigenstate solution of the Dirac equation the mean orbital radius is inversely proportional to the experienced charge:

$$r_{scr} = \frac{a_0}{Z_{scr}} \tag{6.3.2}$$

where a_0=52.9 pm is the Bohr radius. Combining the above two equations, we may calculate r_{scr}:

$$\frac{a_0}{r_{scr}} = Z_{ex}e^{-r_{scr}/r_{TF}}\left(1+\frac{r_{scr}}{r_{TF}}\right) \tag{6.3.3}$$

The above equation gives two bounded solutions for r_{scr}, which we observe by numerical computation. The first one is very close to the ordinary Bohr orbit value, and we interpret it as ordinary bound electron orbitals. The second solution is a surprisingly large orbital radius: the electron feels only a fraction of the positive ion charge from this screened orbital.

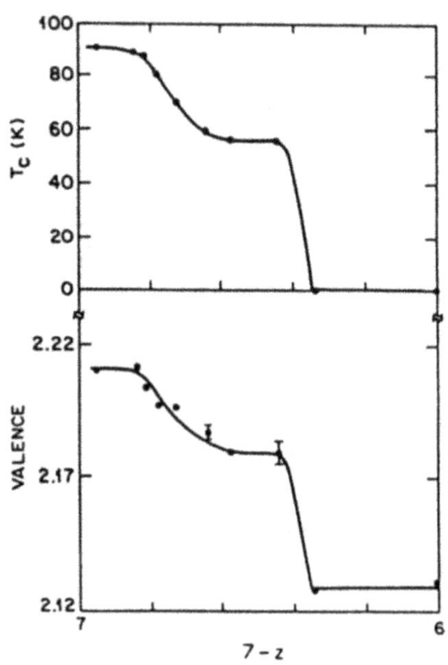

FIGURE 6.3.1. The increase of average copper valence value with growing hole doping rate. Hole doping grows as $z \to 0$ in $YBa_2Cu_3O_{7-z}$. There is a close correlation between the T_c of $YBa_2Cu_3O_{7-z}$ and the valence of its Cu sites, reproduced from [**1**].

Equation 6.3.3 is a first approximation, which calculates the charge that an electron sees from its mean orbit radius. This calculation assumes that the screened s-orbital has the same radial distribution as an ordinary s-orbital, while in reality the Dirac equation's eigenstate will be somewhat different because the experienced charge has a radial dependence. Nevertheless, the simple equation 6.3.3 can be used for estimating the radius, and it illustrates the principles involved.

Let us take the example of $YBa_2Cu_3O_{7-\delta}$ material, where we estimated r_{TF}=0.52 nm. Most Cu sites have approximately $Z = 2$ charge value, as illustrated in figure 6.3.1. This usual Z value appears periodically in the lattice, and generates the usual flat Fermi sea. However, the second solution of equation 6.3.3 must relate to an exceptional Z value, that is then screened over a relatively large distance. As can be seen in figure 6.3.1, $YBa_2Cu_3O_{7-\delta}$ hole doping increases the average Cu valence value. Some Cu atoms thus obtain +3 valence state. Therefore, we may take $Z_{ex} = 1$, and then the second solution of equation 6.3.3 yields r_{scr}=3.1 nm. Such mean radius parameter corresponds to a binding energy of just E_b=3.9 meV. Orbits with such low binding energy can not be realized at room temperature. These estimated r_{scr} and E_b values are quite similar in other hole-doped cuprate materials as well.

The phase transition into weakly bound state happens mostly around the $T = E_b/k_b$ temperature, where the temperature fluctuations become low enough for the weakly bound electron state to exist.

The periodically occurring Cu sites with +1 excess charge create a periodic pattern, comprising the electron density variation illustrated in figure 6.2.1. Such a periodic pattern is the CDW phenomenon. In the CDW context, the above calculated E_b value can be identified with the so-called "charge excitation energy gap" - a term introduced by CDW researchers: E_b is the energy needed for changing the charge value on a lattice site. The extent to which Cu sites with +1 excess charge distort the crystal lattice depends on the presence or absence of weakly bound electrons which counter-balance the +1 excess charge. It follows that the $T = E_b/k_b$ temperature determines the CDW phase transition temperature.

6.4. Effective electron mass considerations

In the preceding sections, we used the mass of an isolated electron in our calculations. As an electron moves through a solid-state lattice, it interacts with the positively charged lattice sites. These interactions generate an effective electron mass, which determines the energy - wavenumber relationship. When this effective electron mass differs from the free electron mass, equation 3.2.3 must be written as:

$$E(k) = E_0 + E_k = E_0 + \frac{(\hbar k)^2}{2m_{eff}}$$

where m_{eff} is the effective electron mass, E_k is the kinetic energy of a delocalized electron with wavenumber k, and E_0 is the lowest energy state in the conduction band. The m_{eff} calculation method can be found for example in [**2**]. Hole-doped cuprate materials are known to have $m_{eff} > m$ [**3, 4, 5**], which must be accounted for. The m_{eff}value applies to both delocalized and weakly bound electrons, as in both cases the electron moves across the lattice. Equations 6.2.6, 6.2.10 and 6.3.3 thus take the following form:

$$\left[r_{TF}^{-1} = k_{TF} = e\sqrt{\left(\frac{\sqrt{m_{eff}}}{\hbar}\right)^3 \frac{1}{\sqrt{2}\varepsilon_0\pi^2}\sqrt[4]{E_F}} \right]_{3d} \qquad (6.4.1)$$

(6.4.2) $$\left[r_{TF}^{-1} = k_{TF} = e\sqrt{\frac{m_{eff}}{\varepsilon_0 \hbar^2}\frac{1}{\pi d_{il}}}\right]_{2d,stack}$$

(6.4.3) $$\frac{a_0}{r_{scr}} = \frac{m_{eff}}{m} Z_{ex} e^{-r_{scr}/r_{TF}} \left(1 + \frac{r_{scr}}{r_{TF}}\right)$$

The $\frac{m_{eff}}{m}$ factor of the above equation indicates that, with all else being equal, the weakly bound orbital radius decreases when the effective electron mass increases.

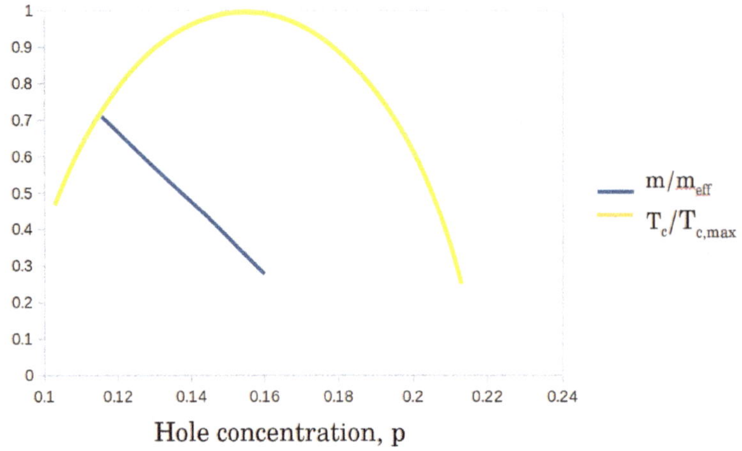

FIGURE 6.4.1. The effective electron mass m_{eff} and superconducting critical temperature T_c as a function of hole concentration. These m_{eff} and T_c data for $YBa_2Cu_3O_{7-\delta}$ are from **[3]**.

Figure 6.4.1 shows how m_{eff} changes with the doping rate in $YBa_2Cu_3O_{7-\delta}$. Since T_c reaches its maximum at the $p = 0.16$ hole concentration, we use the corresponding $m_{eff} \approx 3m_e$ effective mass value for $T_{c,max}$ estimation.

Employing equations 6.4.2-6.4.3, table 1 shows weakly bound electrons' binding energy and phase transition temperature in the representative $YBa_2Cu_3O_{7-\delta}$ superconductor. The obtained phase transition temperature is surprisingly close to the highest T_c in hole-doped cuprates.

Superconductor	Z_{ex}	d_{il}	**Effective e^- mass**	r_{scr}	E_b	E_b/k_b
$YBa_2Cu_3O_{7-\delta}$	$+1$	0.585 nm	$m_{eff} = 3m_e$ $(p = 0.16)$	2 nm	9.1 meV	106 K

TABLE 1. Weakly bound electrons' binding energy and phase transition temperature in the representative $YBa_2Cu_3O_{7-\delta}$ superconductor, calculated from equations 6.4.2-6.4.3.

The highest T_c of $YBa_2Cu_3O_{7-\delta}$ is $T_{c,max}$=93 K near $p = 0.16$ hole concentration. If weakly bound electrons catalyze its superconductivity, then according to table 1 it may still have a potential for reaching higher T_c value. This potential for a higher T_c value in $YBa_2Cu_3O_{7-\delta}$ is validated by reference **[6]**, whose authors found two types of additives that increase the T_c of $YBa_2Cu_3O_{7-\delta}$ to over 100 K.

Figure 5.3.6 shows the weaker $YBa_2Cu_3O_{7-\delta}$ superconducting dome peaking near $p = 0.09$ hole concentration, where $m_{eff} \approx m_e$ can be estimated by interpolating figure 6.4.1 data. For this regime, the previously calculated E_b=3.9 meV value corresponds to $E_b/k_b \approx T_{c,max}$=45 K, which also reasonably matches the experimental $T_{c,max}$ value of this doping region.

Our results explain why large m_{eff} values increase the T_c temperature potential of hole-doped cuprates. The large m_{eff} pre-condition has been noticed already in the past [4], but the underlying physical process was not understood previously.

6.5. The catalytic effect of weakly bound electrons

Up to now, we studied a single electron's weakly bound state. When a weakly bound orbital is occupied by an electron pair, thermal fluctuations shall easily flip it into a spin-triplet state, which is the established pre-condition for their Bose-Einstein condensation.

At the above-calculated E_b/k_b temperature range, even a single electron has a limited lifetime in a weakly bound orbital. When two electrons occupy such an orbital, their bound lifetime is even shorter because their presence causes additional screening of the excess positive charge; they soon transition into delocalized state.

In this sense the continuous transition between delocalized and weakly bound electron states is a catalytic mechanism, which increases the fraction of spin-triplet state delocalized electrons. We propose that this catalytic mechanism explains weakly bound electrons' relevance to superconductivity.

In summary, our weakly bound electron state calculation is based on straightforward energy minimization methodology. The low spin pairing energy between weakly bound electrons explains how spin-triplet electron pairs are produced. Insofar as such electron states catalyze superconductivity, candidate materials' Bose-Einstein condensation temperature must be evaluated according to equation 3.5.11, and their phase transition temperature must be estimated according to equations 6.4.2-6.4.3.

Bibliography

[1] H. Asano "Crystal Structure of High-Tc Superconductors", Nihon Kessho Gakkaishi, Volume 36.2 (1994)

[2] A. Shnirman "Solid state theory I" (2019)

[3] B. J. Ramshaw et al "Quasiparticle mass enhancement approaching optimal doping in a high-T_c superconductor", Science, Volume 348 (2015)

[4] M. Rabinowitz et al "Phenomenological theory of cuprate superconductivity", Applied physics letters, Volume 63.7 (1993)

[5] B. Vignolle et al "Quantum oscillations in an overdoped high-T_c superconductor", Nature, Volume 455.7215 (2008)

[6] M. Z. Gaffoor et al "Investigating the critical transitional temperature increase in graphene oxide doped bulk YBCO", Results in Physics, Volume 44 (2023)

CHAPTER 7

Can inner-shell electrons Bose-Einstein condense?

Andras Kovacs[1] and **Chris Scott**[1]
[1] ExaFuse
E-mail: chris@exafuse.com.au

Abstract. This chapter describes new experimental results that show clear evidences of altered inner-shell electron state formation. From chemistry point of view, these phenomena open up a new field that deals with the manipulation of inner-shell electron states. From physics point of view, these phenomena are unexpected and do not match the known set of electron orbitals. Surprisingly, the measured experimental data are compatible with Bose-Einstein condensation phenomena. Therefore, the intriguing subject matter of this chapter might be a pioneering observation of Bose-Einstein condensed inner-shell electron states.

7.1. Introduction

In preceding chapters, we investigated the Bose-Einstein condensation of delocalized electrons, that occupy the conduction band. The Bose-Einstein condensation process calculations are simpler for delocalized electrons than for bound electrons. However, the explained concepts do not restrict the Bose-Einstein condensation process to delocalized electrons only; such condensation is applicable to any electron state when it becomes thermodynamically favorable. This raises the question of whether bound electrons would Bose-Einstein condense under certain experimental conditions. We pointed out in chapter 1 that high pressure conditions would be thermodynamically favorable, but they are the least interesting for practical applications.

If bound electrons' Bose-Einstein condensation was encountered under ambient conditions, how would one recognize it? Chapter 3 established the thermodynamically balanced co-existence that Fermi-Dirac and Bose-Einstein condensed electron states. In the context of bound electrons, Bose-Einstein condensation is most energetically favorable in the inner-shell electron states, where the energy gain of multi-electron occupancy is measured in kilo electron-volts or even tens of kilo electron-volts. The co-existing electron states would comprise Bose-Einstein condensed inner-shell electron states, surrounded by outer-shell electrons in usual Fermi-Dirac states. To recognize such electron states, inner-shell electron probes must be used. Our measurements involve such inner-shell probes, based on x-ray spectroscopy. The following sections describe three complementing experimental signatures of inner-shell electron states that are occupied by a varying number of electrons. While one should not exclude alternative explanations of these results, our intepretation of Bose-Einstein condensed inner-shell electron states matches all experimental results and fits the previously established theory. It thus becomes the simplest explanation of observed phenomena, favored by Occam's razor principle.

Regardless of the applicable inner-shell electron interaction model, the following experiments demonstrate that the scope of chemistry can be extended into a whole new domain: the manipulation of inner-shell electron states.

In the following, we focus on the explanation of our measurements, without disclosing the details of the involved materials' preparation. Those readers who become interested in this topic are welcome to contact the authors.

7.2. Elevated electron concentration in the inner-shell region

Two samples of a thorium- and bromine-containing material, each weighing 0.2 g, are placed into the anodic and cathodic compartments of a CR2032 type coin cell. These samples comprise a solid state material with rubber-like consistency, and have sufficient electronic and ionic conductivity for the applied electro-chemical process. The two coin cell compartments are separated by a cation-conducting membrane. A charging current of 0.5 mA is passed through the cell for 20 hours, i.e. electrons are pushed into the anodic compartment. Thereafter, the sample is taken out from the anodic compartment, and its x-ray emissions are monitored by passive x-ray spectroscopy technique. Figure 7.2.1 shows the obtained x-ray spectra. The observed inner-shell excitation peaks are higher during the initial 2 hours than in the subsequent 24 hour measurement. During the initial 2 hours, even the Po L-α peak is well discernible, while it fades into noise during later measurement. The lowest energy region of the x-ray spectrum is also higher during the initial 2 hours than in the subsequent 24 hour measurement, and this low-energy excess is consistent with the braking radiation of energized electrons.

In contrast, performing the same measurements on the sample from the cathodic compartment yields static x-ray spectra. In both compartments, the electron-migration and ion-migration processes are the same. The difference is the introduction of excess electrons in the anodic compartment versus bromine oxidation in the cathodic compartment.

This data demonstrates that some inner-shell electron states can be re-arranged by the introduction of excess electrons, yielding a metastable configuration. More specifically, figure 7.2.1 shows the largest percentage-wise changes for the Tl, Bi, and Po peaks, which are at the end of the thorium decay chain. The nuclear decays of ^{212}Pb and ^{212}Bi emit energetic particles, which may ionize L-shell electrons, and the observed Tl, Bi, and Po peaks correspond to x-ray emission during subsequent re-fillings of the L-shell. At the start of the thorium decay chain, the nuclear decay rate remains constant. This constant nuclear decay rate can be seen in the Ra peaks' static amplitudes, which correspond to the $^{232}Th \rightarrow {}^{228}Ra + {}^{4}He$ and $^{228}Th \rightarrow {}^{224}Ra + {}^{4}He$ decay events. Assuming a constant or approximately constant nuclear decay rate, the strongly elevated Tl, Bi, and Po peaks indicate a strongly increased electron concentration in the inner-shell region of involved atoms; such an increased electron concentration proportionally increases the L-shell ionization probability. A key point: increasing the electron concentration in an inner-shell orbital is possible only with Bose-Einstein condensed inner-shell electron states. Furthermore, it is anticipated that the introduction of excess electrons would increase the electron occupancy of Bose-Einstein condensed states because some incoming electrons fall down into the inner-shell Bose-Einstein condensate. The rising inner-shell peaks of the anodic compartment can be interpreted as this anticipated inner-shell electron density increase.

Nuclear decays emit highly energetic electrons or ^{4}He particles. When a highly energetic particle collides with Bose-Einstein condensed electrons, it breaks up these electrons' phase coherence. Consequently, the vacant inner-shell electron states become re-filled with incoherent electrons, emitting x-rays at the usual de-excitation energy levels. In this case, a larger number of electrons are ionized from the inner-shell than the number of electrons re-filling these states. Such an excess of ionized electrons produces excess braking radiation, which explains the elevated x-ray readings in the low-energy part of the spectrum. The excessive ionized electrons may also knock out electrons from nearby Br and Th

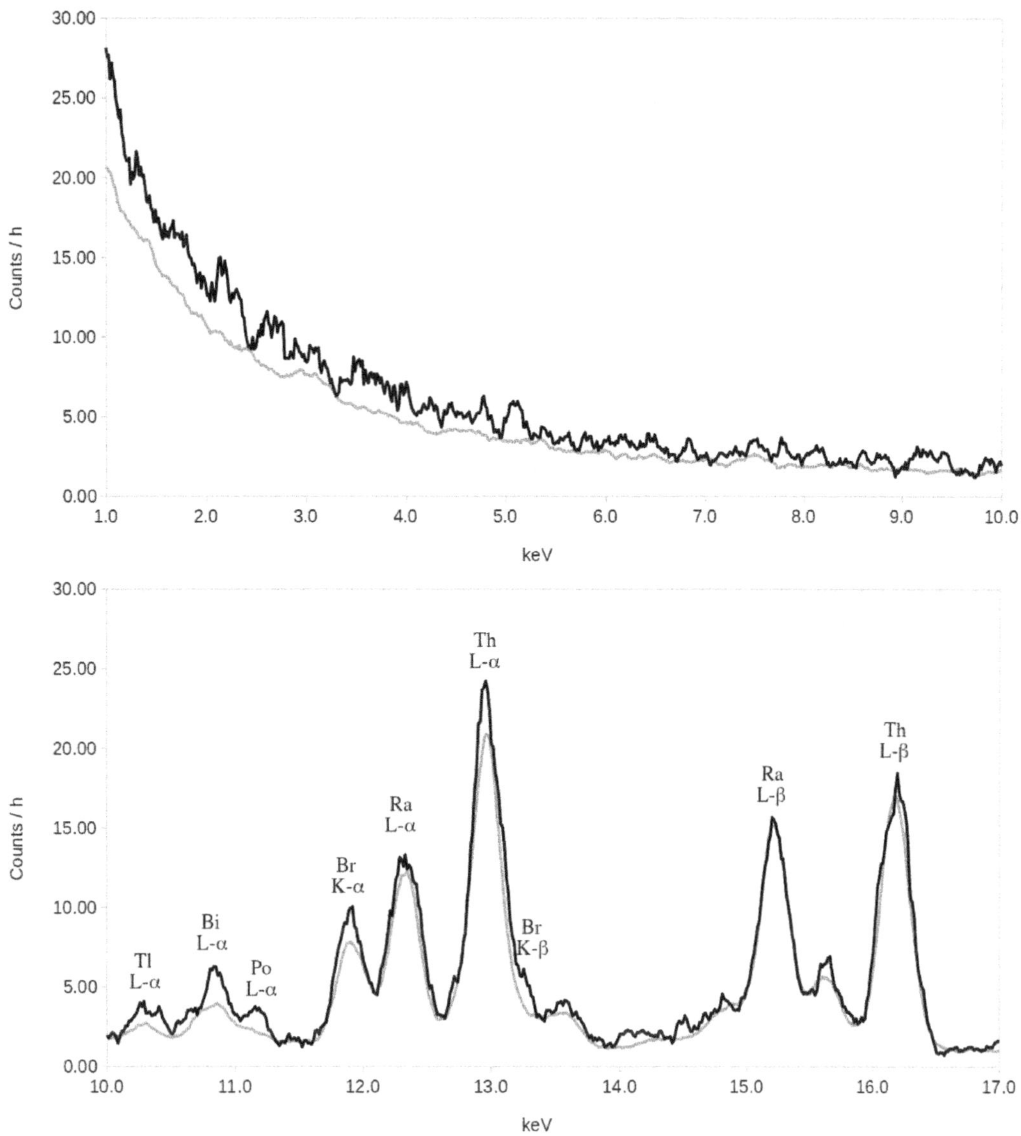

FIGURE 7.2.1. Overlayed x-ray spectrum measurements on a thorium- and bromine-containing material, that was electro-chemically processed. Black colour: measured shortly after the electro-chemical procedure, the measuring time is 2 hours. Gray colour: next measurement, the measuring time is 24 hours.

atoms, resulting in a slight increase of Br and Th peaks, that are also visible in figure 7.2.1.

In summary, the observed phenomenon is compatible with Bose-Einstein condensed inner-shell electron states. It is difficult to find an alternative explanation for the strongly elevated Tl, Bi, and Po L-α peaks.

7.3. Magnetically induced x-ray emission

For this experiment, we use the same type of thorium- and bromine-containing material that was used in section 7.2. Before the experiment, it is saturated by electrons. A neodymium magnet is placed onto a sample, which is already on the x-ray spectrometer. A few seconds later, we start monitoring the sample material by passive x-ray spectroscopy. This procedure ensures that the x-ray spectrometer operates under a static magnetic field,

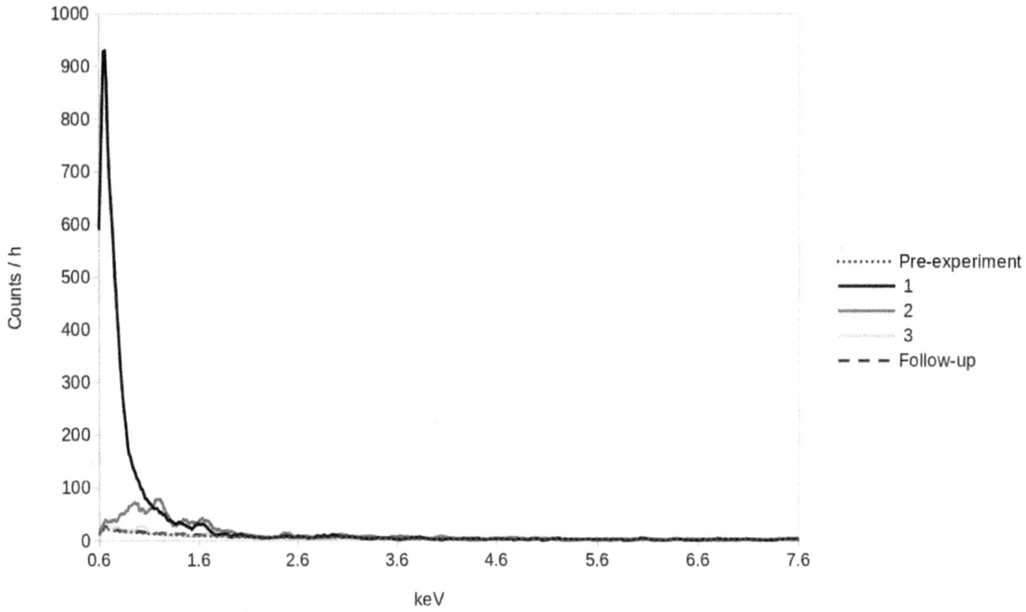

FIGURE 7.3.1. The x-ray spectrum of a magnetically induced x-ray burst, emitted by the thorium- and bromine-containing material. The chart shows consecutive x-ray spectra, saved at 15 minute intervals, as well as long duration pre-experiment and follow-up spectra. All measurements are normalized to the same measurement duration, and were obtained by passive x-ray spectroscopy method. Pre-experiment: measured just before the experiment, 36 hours duration. 1: a 15 minutes measurement that starts directly after a magnet is placed onto the sample. The magnet remains on the sample during the following measurements. 2: next 15 minutes measurement. 3: next 15 minutes measurement. Follow-up: a 2 hours measurement that starts directly after the previous measurement.

which does not influence the instrument operation[1]. Figure 7.3.1 shows the consecutive x-ray spectra measurements, saved at 15 minute intervals, as well as the pre-experiment and follow-up spectra. It can be seen in figure 7.3.1 that the applied magnetic field triggers a relatively strong x-ray emission in the 1-2 keV energy range. This emission lasts for 10-15 minutes, and its energy extends up to 2.2 keV. Figure 7.3.2 shows the same data as figure 7.3.1, but represented on a logarithmic scale. In the lowest energy range of 0.5-1 keV, the attenuation of pre-experiment and post-experiment spectra is analogous to the attenuation of the burst signal. The observed x-ray emission cannot originate from any ordinary chemical processes, but only from sufficiently energetic inner-shell electrons.

We established in the preceding chapters that magnetic fields are detrimental to electrons' Bose-Einstein condensed state. In chapter 2, we investigated the physical process behind superconductors' perfect diamagnetism. Bose-Einstein condensed inner-shell electrons shall analogously respond to an applied magnetic field; it shifts the thermodynamic balance between the Fermi-Dirac and Bose-Einstein condensed electron states. The resulting re-arrangement of some inner-shell electrons involves a large change in their binding energies. We therefore propose that magnetically induced x-ray emission is a consequence of the magnetically induced disruption of some inner-shell electrons' Bose-Einstein condensed state.

[1] We verified by measurements that our x-ray spectrometer is not influenced by static magnetic fields.

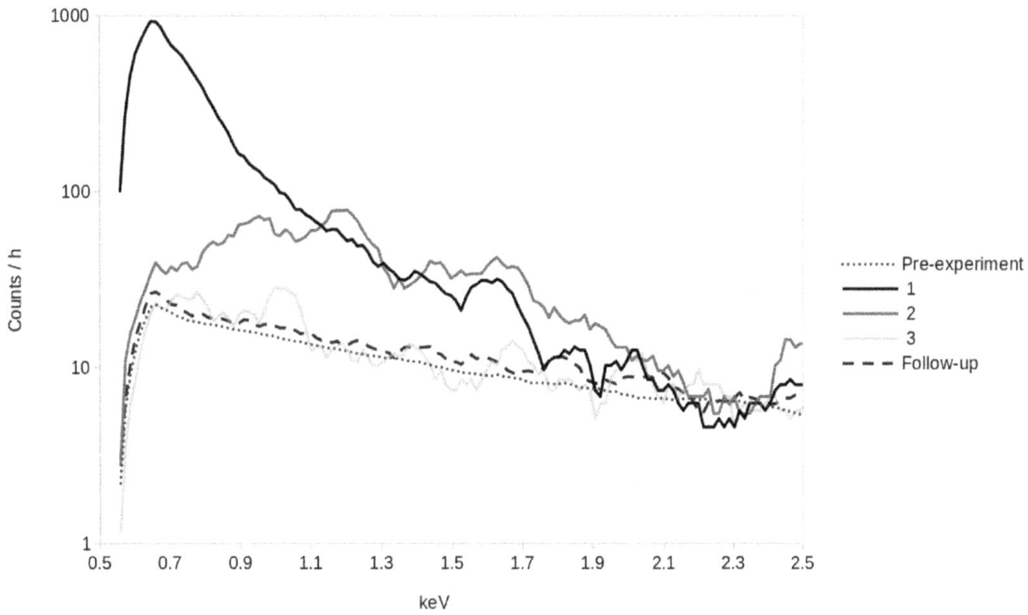

FIGURE 7.3.2. The same data as figure 7.3.1, but represented on a logarithmic scale, and showing the spectrum only up to 2.5 keV. Here, the low-energy attenuation region can be well observed.

7.4. Anomalous inner-shell electron excitation lines

Suppose that N number of Bose-Einstein condensed electrons occupy the inner-most K-shell orbital. Taking any one of these electrons, statistically we find $\frac{N-1}{2}$ electrons that are radially closer to the nucleus, and $\frac{N-1}{2}$ electrons that are radially further away from the nucleus. These Bose-Einstein condensed electrons therefore perceive an effective nuclear charge of $Z - \frac{N-1}{2}$, where Z is the actual nuclear charge. As a consequence, the electron binding energy of the inner-most orbital is different for Fermi-Dirac versus Bose-Einstein condensed electron cases. With an appropriate measurement method, which does not break up such a Bose-Einstein condensate, it should be possible to measure this energy difference.

7.4.1. Anomalous electron excitation measurement on a pre-treated copper surface. A material comprising mainly copper is prepared in thin film form. While remaining under an inert atmosphere, the surface of this material is illuminated for a month by a green laser pointer light. This laser illumination causes a surface color change. Subsequently, this sample is analyzed by XRF technique; the obtained spectrum is shown in figure 7.4.1. This spectrum contains an unexpected peak appearing at 5.44 keV, which is close to the K-α peak of Cr. However, while this XRF peak is prominent at 8 kV XRF voltage, it is barely visible in the 15 kV spectrum, and it is completely absent in the 40 kV and 50 kV spectra. Such anomalous peak counts are in contrast to the peak count evolution of physically present elements, such as Ar (ambient) or Cu, whose XRF peaks are also in this x-ray energy region. Therefore, the XRF peak appearing at 5.44 keV does not correspond to any ordinary inner-shell state; it is a signature of an unconventional inner-shell electron state. This unconventional inner-shell electron state is further characterized by the following anomalous features of this 5.44 keV peak: its peak energy is 25 eV higher than a real 5.415 keV centered Cr K-α peak, it is wider than the XRF peaks of ordinary inner-shell electron states, and its shape is asymmetric with respect to the center-line at 5.44 keV energy. We also note that our sample material does not have any Cr contamination.

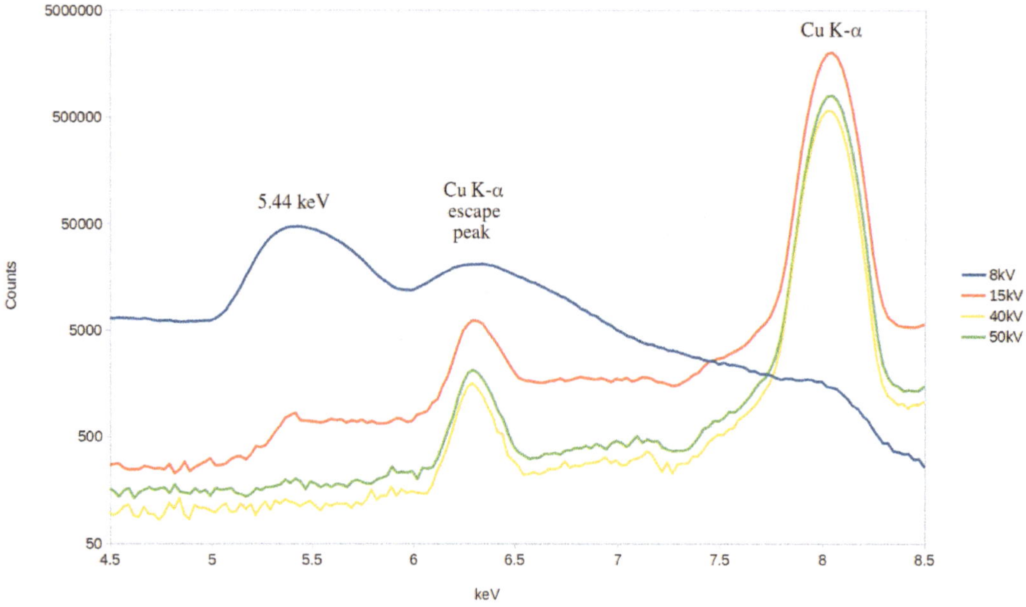

FIGURE 7.4.1. XRF spectrum measurements on a copper-based material after laser illumination. The labels indicated the applied voltage in the XRF device.

The same sample was also analyzed by EDS technique, using 30 keV impacting electron energy. The x-ray spectra obtained by EDS do not contain any peak near 5.44 keV. With reference to section 7.2, such impact of such high energy electrons breaks up any Bose-Einstein condensed electron states.

The anomalous 5.44 keV peak cannot be a Bragg diffraction peak because the angle between the x-ray source and detector is almost 360°, which is well outside the Bragg diffraction condition. Furthermore, reference [1] demonstrates that if a Bragg diffraction peak shows up under one XRF voltage setting, it shows up at the same energy position under all other XRF voltage settings as well.

This anomalous XRF peak can be interpreted on the basis of above explained effective nuclear charge, perceived by Bose-Einstein condensed inner-shell electrons. I.e. the 5.44 keV XRF peak demonstrates that in a fraction of atoms, inner-shell electrons are bound to an "effective nuclear charge" of +24, which is lower than the actually present Z=29 charge value of Cu nuclei. The Z=29 and $Z - \frac{N-1}{2}$=24 values mean that, on average, N=11 electrons occupy the inner-most K-shell. We do not observe any change in our sample material with the passage of time, i.e. this electron configuration appears to be stable under ambient conditions.

7.4.2. Anomalous electron excitation measurement on a pre-treated quartz surface. In this experiment, we operate a krypton-filled quartz tube, with an electrically heated filament in the middle. The filament is powered by AC current, and its temperature reaches around 2500°C. To avoid overheating, we operate the krypton-filled tube under repeating ON-OFF regime. A krypton-filled tube was operated in this way for 8 days, and remains operational at the conclusion of the experiment run.

After the experiment, we notice a bump-like spot on the quartz surface, and it has opaque gray coloration. In repeated experiment runs, the appearance of such a spot is reproducible with good probability.

We break the tube open after the experiment, and analyze the inner surface of the gray spot by XRF technique; the obtained spectrum is shown in figure 7.4.2. This spectrum

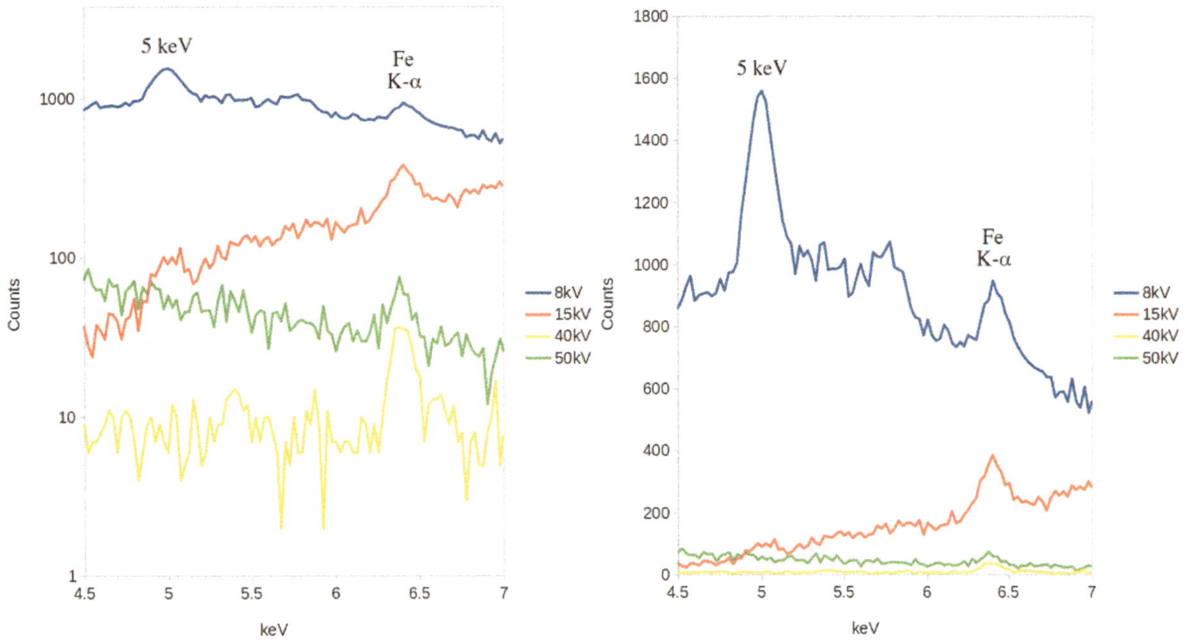

FIGURE 7.4.2. XRF spectrum measurements on the gray spot that appears on quartz surface after the experiment run. Logarithmic scale is shown on the left, and linear scale is shown on the right. The labels indicated the applied voltage in the XRF device.

contains an unexpected peak appearing at 5 keV, which is close to the K-α peak of vanadium. As in the previous example, this XRF peak is prominent at 8 kV XRF voltage, barely visible in the 15 kV spectrum, and completely absent in the 40 kV and 50 kV spectra. Such anomalous peak counts are in contrast to the peak count evolution of physically present elements, such as Fe, whose XRF peak is also in this x-ray energy region. Therefore, the XRF peak appearing at 5 keV does not correspond to any ordinary inner-shell state; it is a signature of an unconventional inner-shell electron state. This unconventional inner-shell state has 48 eV higher peak energy than a real 4.952 keV centered V K-α peak. We also verified by XRF that the employed filament material does not have any vanadium contamination.

As before, we interpret this anomalous XRF peak on the basis of effective nuclear charge, perceived by Bose-Einstein condensed inner-shell electrons. The 5 keV XRF peak demonstrates that in a fraction of atoms, inner-shell electrons are bound to an "effective nuclear charge" of +23, which is lower than the actually present Z=36 charge value of Kr nuclei. The Z=36 and $Z-\frac{N-1}{2}$=23 values mean that, on average, N=27 electrons occupy the inner-most K-shell. We do not observe any change in our sample material with the passage of time, i.e. this electron configuration also appears to be stable under ambient conditions.

Lastly, we note that the operating condition of Kr gas is reminiscent of the fast-cooled palladium-deuteride example from chapter 3, which retains its thermally generated spin-triplets during fast cooling, and thus becomes superconducting already at 60 K temperature. Within the krypton-filled tube, Kr atoms heat up to 2500°C temperature at the tungsten surface, while also being exposed to intense light radiation in the few eV photon energy range. Krypton's orbital electrons therefore have a much higher probability of spin-triplet formation at the tungsten surface than under ambient conditions. Some spin-triplets comprising Kr atoms quickly cool to near-ambient temperature when they reach the quartz surface; the probability of such spin-triplet electron pairs' Bose-Einstein condensation is therefore maximized at the inner quartz surface, which captures the resulting pseudo-atom. In summary, our experimental observations can be interpreted according to the already established electron Bose-Einstein condensation principles.

Acknowledgements: The authors thank Heikki Sipilä for useful advices and consultation. The XRF measurements have been performed at the FinFocus company.

References. [1] R. Tanaka et al "Artificial peaks in energy dispersive X-ray spectra: sum peaks, escape peaks, and diffraction peaks", X-Ray Spectrometry, volume 46.1 (2017)

www.ingramcontent.com/pod-product-compliance
Lightning Source LLC
Chambersburg PA
CBHW082347230926
51846CB00004B/942